Smithsonian Intimate Guide to the Cosmos

Smithsonian | Intimate Guide to the Cosmos

Visualizing the New Realities of Space

DANA BERRY

Introduction by Eric J. Chaisson

A GREYSTONE / Madison Press Book

GREYSTONE BOOKS

Douglas & McIntyre Publishing Group

Vancouver / Toronto

Printed in Singapore

Published in Canada in 2005
by Greystone Books
A division of Douglas &
McIntyre Ltd.
www.greystonebooks.com

Library and Archives Canada
Cataloguing in Publication data
is available on request

ISBN 1-55365-080-8

(Above) Horsehead
Nebula, IC434.
(Previous page) Star-forming
Region SI06IRS4.

Produced by
Madison Press Books
1000 Yonge Street, Suite 200
Toronto, Ontario
Canada M4W 2K2

Contents

*This book is
dedicated to
Zeb Vance Berry Sr.
and his grandson,
Joshua Zane Berry*

6 **Introduction**
by Eric J. Chaisson

10 **Preface**

14 **Chapter One**
Our Place among the Stars

20 **Chapter Two**
Earth and Environs

56 **Chapter Three**
Is There Life in Outer Space?

74 **Chapter Four**
Other Planets, Other Suns

108 **Chapter Five**
The Milky Way and Beyond…

126 **Chapter Six**
The Starry Messengers

148 **Chapter Seven**
The Big Picture

164 **Epilogue**

170 Index and Credits

by Eric J. Chaisson

That we live in a visual society today is an understatement. We are bombarded with images of all sorts — from television to advertisement to the World Wide Web — that seemingly engulf us at home, work and play. Not all of it good stuff, and not all of it wanted. Yet there's something in our psyche that attracts us to pictures, patterns, art and animation. Music for the eyes. Conveyance via film. Invisible made visible. Visualization takes us vicariously to places — both real and imagined — that we often never actually attain, for our mind's eye widens with imagery like no other media.

That said, we also like to talk, read and write. Verbal expression was a hallmark in the evolution of our hominid forebears for well longer than visual symbolism. Even today, though aided by technology, there is perhaps no more effective learning device than a good book and a gifted lecturer. A thousand words may equal a single picture, but a carefully crafted thousand words can render a remarkable view. Words, too, feed our imaginations, enhance our curiosity and transport us to locales far and foreign.

At the dawn of the new millennium, our fast-paced society teeters on the brink of a cultural crisis. Will it be words or pictures? Books or monitors? Adherence to grammar and syntax, or anything-goes visual impressionism? How shall we communicate in a technological democracy that relies on information content, delivered accurately and effectively?

"This book is an uplifting work of descriptive imagery tha

The enterprise of science, if nothing else in our subjective society, requires that we champion objectivity. There is no alternative, for that is what distinguishes the scientific method from other notable endeavors. Yet when it comes to disseminating good, solid, accurate science, especially in astronomy, the subject matter often benefits from not only artistic license in imagery but also carefully chosen words to make that captured science real and understood.

Every once in a while, someone comes along who skillfully combines the two. The visual and the verbal are synthesized. The art and imagery soar, and the accompanying words and phrases do too. The eyes dominate the work, the hands turn the pages, and the brain soaks it up — scanning the illustrations, reading the text, all the while becoming enthralled as our curiosity peaks. That eye-hand-brain amalgam, whose exquisite qualities more than any others made us who we are, has a field day.

This volume is one such product — a stunning blend of words and images, of text and art. Written and illustrated by one of America's foremost space artists, this collection truly *does* take us to the far away and long ago.

Dana Berry has emerged as one of today's finest visual interpreters of science. Combining the pizzazz of an artisan with the knowledge of a scientist, Dana brings to his work unmatched insight and technical accuracy. And can he write! For

closely approximates technological poetry." — Eric J. Chaisson

eleven years, I had the pleasure of working alongside Dana, often marveling at his doodlings sketched on luncheon tablecloths, airbrushed paintings done on his kitchen counter, and dazzling animations created with some of the world's most powerful workstations. But I never knew he could write so well.

The result is indeed a synthesis of the visually spectacular and the verbally compelling. Here, Dana takes us on a tour of the cosmos — from our home at planet Earth to the limits of the observable Universe — distilling, through artistic expression of deep space astrophysics and creative packaging of downlinked images from robotic spacecraft, his particular brand of science communication. The work is unparalleled.

We need not oversell human exploration for its economic impacts, nor hype today's astronomy for its practical applications. There aren't any, or many. It's enough that we seek knowledge for the sake of knowledge, witness beauty for the sake of beauty, perhaps achieve a measure of truth while unlocking secrets of the Universe. Dana Berry's new book is a testimonial to how humanity does that — an uplifting work of descriptive imagery that closely approximates technological poetry.

— *Eric J. Chaisson*

Up

beyond the clouds and shouting wind, beyond the edge of the sky, lies the cosmo infinite and mysterious, beckoning to us, challenging our notions of who and what we are.

My own adventures with the cosmos began when I was a kid and my dad gave me a small telescope. We used it mainly to look at the Moon, since that was in those heady days of the *Apollo* Moon landings. Night after night we set up in the driveway of our home in rural South Carolina and watched the Moon work its way through all its phases. We could hardly peer through the eyepiece for more than a few minutes before we suffered the red splotches of retina burns.

It still amazes me that light from the Sun reflected off the Moon so brightly, illuminating the surrounding yard and fields, that we didn't even need a flashlight. Thinking of that now takes me back to those warm summer nights with crickets buzzing and frogs croaking, the sweet smell of magnolias and a fat, full Moon, where two astronauts were walking around at that very moment, looking back at us.

My dad was always perplexed by the heavens. Though not a religious man, he one day pointed out to me that the first book of the Bible began with a story about the origins of the cosmos. Why should that be, he asked? *"In the beginning, God created the heaven and the earth. And the earth was without form, and void; and darkness was upon the face of the deep."* These lines speak to me more about the authors of Genesis than they do about God or the cosmos: our cosmic origins mattered so greatly to the ancients that they opened their holiest book with a profound statement on the subject. Those origins matter no less to us today — perhaps even more so — and my father's question continues to haunt me.

This book will explore and celebrate some of the changes in our understanding of the cosmos. When the last astronauts left the Moon, they imbued a new generation of scientists and explorers with a sense of optimism and hope. Decades later, that generation has discovered a dazzling array of new wonders in the cosmos. And astronomy as we knew it in the days of *Apollo* — as I knew it from childhood — has changed radically.

Welcome to the *new* cosmos.

Earth as seen from *Apollo 17*, then en route to the Moon on December 7, 1972. No human being has been this far away from Earth since the completion of the *Apollo 17* mission.

CHAPTER ONE

Our Place among the Stars

From our home on Earth, we look out into the distances
and strive to imagine the sort of world into which we were
born. With increasing distance our knowledge fades until
at the last dim horizon we search among ghostly errors
for landmarks scarcely more substantial. The search will
continue. The urge is older than history. It is not satisfied
and it will not be suppressed.

— Edwin P. Hubble, from *Realm of the Nebulae*, 1936

We are living in a new golden age of astronomy, greater than
that inaugurated by Galileo. It is an era in which many fundamental questions are
being seriously reconsidered. Some of these are perennial questions: "Where did
we come from? How did the Universe begin? Is there life in outer space? What is
the fate of the Universe?" Others we didn't know to ask before: "What are dark
matter and dark energy? Has time always been the same? What is a black hole?"

We have now begun to gather some of the answers to these mysteries.
Astronomers, historically concerned only with the motions of celestial bodies and
cosmology, are today informed by physics, biology, chemistry, geology, meteorology,
cartography, statistics, optics and computer science. New tools have been developed
to aid them in their quest for cosmic origins. Small rovers explore the surface of
Mars, while probes orbit Venus and Jupiter. We've touched down on an asteroid,
charted the Moon and are monitoring the Sun and its connection to our home planet.
Powerful new telescopes in orbit around Earth (such as the *Hubble Space Telescope*
and the *Chandra X-ray Observatory*), along with technical innovations for ground-
based observatories, are allowing us to probe the cosmos at all scales.

We cannot forget, however, the high price we sometimes pay for these great advances. The loss of the crew of the space shuttle *Columbia* underscores the dangers of spaceflight. Yet, even in the shadow of tragedy, our knowledge continues to grow. In the aftermath of the *Columbia* disaster, when people were questioning the validity of the space program, it was easy to overlook one of the most startling advances made to date in our understanding of the heavens. On February 11, 2003 — only ten days after *Columbia*'s loss — the results from a little-known space probe were released. The news was so big that nothing like it has been heard since the day the world was proven round. As one astrophysicist put it, "Astronomers will remember what they were doing when they heard [about it]."

Chuck Bennett and the team running the *Wilkinson Microwave Anisotropy Probe* (*WMAP*) announced a series of findings that were unthinkable only the day before. From this data, we now know that the Universe is 13.7 billion years old — give or take a hundred million years. The composition of the Universe is four percent atoms and matter we would recognize; twenty-three percent mysterious dark matter; and seventy-three percent dark energy, a form of energy that was generally unkown before 1997. This means the Universe will expand forever, and will not end in a "big crunch," as some have speculated.

Bennett and his team confirmed that the Universe is in a new period of accelerated inflation: not only is it expanding, but its rate of expansion is also speeding up. *WMAP* also shows that the first stars "turned on" approximately 200 million years after the initial Big Bang — far, far earlier than astronomers thought. This suggests that stars began forming long before most galaxies were fully developed.

All of this comes from a single, tie-dyed image that charts anomalies and slight temperature fluctuations in the microwave background. The microwaves seen in *WMAP*'s image are the same as those generated by a household kitchen oven — only *WMAP*'s microwaves are considerably older, and ubiquitous throughout the sky. In fact, the microwaves recorded by *WMAP* are not just old, they are the fading embers of the Big Bang itself. Microwaves are a form of electromagnetic radiation, just like radio waves, X-rays and light. Our eyes can't see microwaves, but then, we can't see radio waves or X-rays either. We must instead rely on technology

The back wall of the Universe as seen by *WMAP*. The splotches in this image are from the oldest light that can be detected — the microwave background radiation. Nothing can be seen beyond this wall of light, which dates from 379,000 years after the Big Bang.

to help us see in other regions of the electromagnetic spectrum.

The fluctuations in *WMAP*'s image show variations of about one millionth of a degree and reveal a mottled pattern thought to be the nascent foundation of large-scale cosmic structure as we see it today. The blues in the image indicate regions where the Universe was slightly cooler, slightly more dense. The reddish splotches show regions where the Universe was warmer, slightly more energetic. The image dates back to a time when the Universe was without structure. It is a snapshot of the cosmos at the moment of decoupling — when the Universe was transformed from a homogeneous, opaque sea of subatomic particles into a clumpy, transparent cosmos characterized by complexity.

All of this information was generated by 840 kg (1,850 lb) of aluminum, composites and plastic floating in space roughly one million miles from Earth. This tiny spacecraft spent a year scanning the entire sky, mapping the minute variations in the microwave background.

Much of the success of *WMAP* was predicated on a mission called the *Cosmic Background Explorer* (*COBE*), launched in 1989. *COBE* was designed to probe that faint echo of the Big Bang first confirmed by Arno Penzias and Robert Wilson at Bell Labs in 1965. The data from *COBE* was tantalizingly consistent with predictions for a Universe that began with a Big Bang, but it was clear that a new mission with greater sensitivity was needed. Hence, *WMAP*.

The Echo of the Big Bang

In 1962, radio astronomers Arno Penzias, left, and Robert Wilson were troubled by static noise in their radio dish at Bell Labs in Holmdel, New Jersey. The noise sounded like radio static but was detected in the microwave part of the electromagnetic spectrum — and it was uniform in all directions. After conducting an initial investigation, they ascertained that it didn't come from another facility or source, nor was it caused by a defect in their equipment. Perhaps there was a theoretical explanation for the noise. They contacted Princeton astronomer Robert Dicke, who at that time was pursuing various theories about the Big Bang. Dicke believed that if there had, in fact, been a Big Bang, the residue of the explosion might now be evident as low-level background "noise" throughout the Universe. Eventually, Penzias, Wilson and Dicke published the results of their research — the discovery of the cosmic microwave background. This was the first direct evidence of the Big Bang since Edwin Hubble's discovery of galactic redshifts. For their work, Penzias and Wilson were awarded the Nobel Prize for Physics in 1978.

And this is just the beginning. *WMAP*'s image shows us the Universe at a very early stage, but that picture is also a boundary. Today, we can't see beyond decoupling. No technology currently lets us see anything more distant, more aged than the microwave background. But that might change soon.

A small flotilla of spacecraft known collectively as *LISA* (*Laser Interferometer Space Antenna*) will in the next decade probe the mysterious realm of gravity waves. Exploration of the gravity spectrum — that range of gravity wave frequencies and wavelengths analogous to the more familiar electromagnetic spectrum of light and radio — is an entirely new field of research. *LISA* will investigate a variety of gravity wave-producing phenomena, most notably colliding black holes and hypernovae, but it might, at the very limits of its capabilities, be able to observe hints of cosmological gravity waves — waves that were caused by the Big Bang itself. If *LISA* is able to detect these waves, it will give us a glimpse of the Universe beyond decoupling, beyond

WMAP. Cosmological gravity waves will show us the Universe at the very moment of its creation.

Closer to home, particle accelerators are providing a laboratory in which to study the behavior of subatomic particles under extreme conditions similar to those of the early Universe. Yet the greatest revelation that has emerged from all of this research and exploration — one that lies at the heart of every scientific endeavor — is that the Universe *is* knowable. And we have now begun to know it.

It was Sigmund Freud who pointed out that scientific revolutions tend to marginalize humans. Copernicus removed our world from the center of God's creation and put our planet in the suburbs of the Sun. Darwin wrenched humanity from its once-divine ancestry. Now *WMAP* has shown us that even the very atoms listed in a periodic table — the chart that confronts every high school chemistry student and catalogues all the basic materials we can see, touch, taste and feel — represent only four percent of what's really there. Slowly we have learned that we are not the center of creation (we may, in fact, be entirely incidental) and that the basic elements of our everyday reality are small-time stuff — just the side effects of star formation.

While we may no longer be the center of a divinely created Universe, our ability to probe and understand the cosmos is quite a consolation. No other species on our planet — perhaps no other species *anywhere* — has this ability.

The last century opened with a debate over the nature of spiral nebulae and a newfangled theory called relativity. Seventy years later, we already had the first direct evidence of the Big Bang, we had split the atom and we had taken our first tentative steps into space. The twentieth century closed with an orbiting space station, a near-comprehensive tour of the solar system, the discovery of a fifth force in nature, an inflationary Universe and awareness of more planets outside our solar system than within. We have begun to know the workings of our planet, its weather, its oceans and its origins. We have even begun to discover the secrets of life itself.

We have learned so much in such a short period of time. It's as if we as human beings are awakening for the first time and are beginning to understand our cosmic connections and context. Slowly, we are taking our place among the stars.

CHAPTER TWO

Earth and Environs

Geographers crowd into the edges of their maps
parts of the world which they do not know about, adding notes
in the margin to the effect that beyond this lies nothing
but sandy deserts full of wild beasts.

— from *Plutarch's Lives*

The Empire of the Sun | It is no accident that astronomy is our oldest science. We needed it for agriculture, to mark time and to predict the seasons. The motions of the stars and planets and the cycles of the Sun and Moon told us when to plant next year's crop and also gave us a way to mark the flow of time and record unique events during those regular cycles.

The ancient Greeks believed that the foundations of the cosmos must somehow be geometrically perfect — and that if the Universe was perfect, it must also be static. But astronomy relies on direct observation, and disturbing evidence kept interfering with this heavenly perfection. While the heavens appeared to be revolving in a timeless fashion, observers noticed odd anomalies. Certain heavenly bodies, the so-called *planaomai* (Greek for "wandering ones"), wouldn't behave. Night after night, the position of Mars, for example, changed — sometimes even seeming to move backward — until it traced out a great loop in the sky.

In order to keep Earth at the center of the cosmos, the ancient astronomer Ptolemy had to devise a tangled scheme of epicycles — complicated, eccentric positions for the planets — that would account for, as well as predict, a planet's position. The trouble was, it failed.

Thanks to a fleet of robotic probes, we have managed to glimpse up close nearly every planet and moon in our solar system. The collage opposite is a group portrait of several of these planets.

The Ptolemaic model of our Universe. The Earth (A) resides in the center of the cosmos. In a tight orbit around the Earth are our Moon, Mercury, Venus, our Sun, Mars, Jupiter, the ringed planet Saturn and the celestial sphere (B–I respectively). Ptolemy, an astronomer who lived in Alexandria (approx. 87–150 A.D.), derived his model from ancient Greek examples, all of which were Earth-centric. His model remained the accepted standard until Copernicus "revolutionized" the Universe 1,400 years later.

No progress was made until Nicolaus Copernicus published his *De Revolutionibus Orbium Celestium* in 1543, shortly before his death. His theory moved the Sun into the center of the solar system and claimed that Earth rotates on its axis once every day and revolves around the Sun once every year. These conclusions enraged church officials, who vigorously disputed the Copernican model. The Polish astronomer's ideas were later refined by the ellipses of Johannes Kepler, the falling apple of Isaac Newton and many others.

Our understanding of the solar system continued to grow steadily, albeit incrementally, until the latter half of the twentieth century when we finally visited some of the far-flung corners of the empire of the Sun.

In the Beginning | So where did our solar system come from?

About four and a half billion years ago, some vast amorphous cloud of gas and/or dust suffered a major disturbance, perhaps the shock wave from a nearby

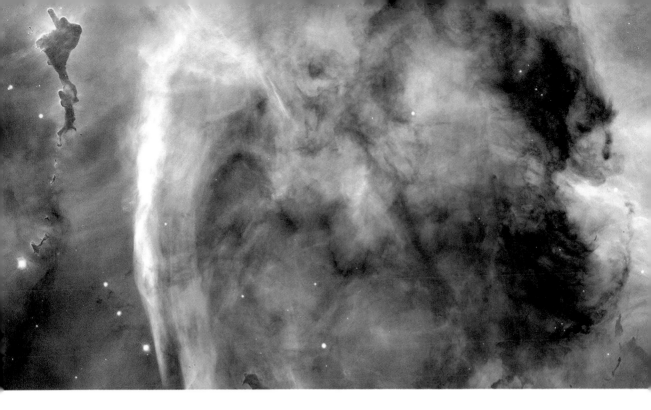

supernova passing through it. That supernova helped seed the cloud with heavy elements — the silicates, metals and heavy gases needed to make worlds. The cloud collapsed under its own gravity and, as it did so, its spin increased — forming a disk and heating up. More and more matter sank to the heart of the disk, increasing its density, temperature and pressure until the fires of atomic fusion ignited. (Fusion is that process in which elements with small atomic nuclei are combined into "heavier" elements with larger atomic nuclei. In the Sun, as in most stars, this is generally the conversion of hydrogen into helium). Once fusion began, solar winds from the newborn star blew away remaining debris, leaving only the protoplanets behind — and the solar system was born.

NGC 3372 — better known as the Key Hole Nebula — is an awesome example of an amorphous cloud that is forming solar systems similar to ours. This image of the Key Hole, roughly 8,000 light-years from Earth, spans about nine light-years from top to bottom — more than twice the distance between our Sun and our nearest stellar neighbor, Proxima Centauri. The image was released from the *Hubble Space Telescope* in 2000.

The nebular theory has long been widely accepted, but it was not until the *Hubble Space Telescope* provided direct observations of the Orion Nebula in the early 1990s that we could see the theory in action. These observations led Robert O'Dell (formerly of Rice University) to the discovery of proplyds — short for protoplanetary disks, spinning clouds of protoplanetary dust circling a warming stellar globe — in that star-forming region. Our own solar system may have been born in a similar star factory.

Other recent evidence supports the validity of the nebular theory. Astronomers have described a diffuse, peanut-shaped bubble of hydrogen gas — thought to be a wispy supernova remnant that might be connected with the collapse of the cloud we came from — that surrounds our star. Some 225 light-years in diameter, this bubble also engulfs all of our nearest neighbors, including the Alpha Centauri star system, Beta Pictoris (a nearby star surrounded by a huge disk of gas and dust similar to the proplyds observed in Orion), and a faint, inconspicuous star named HD 44594. (Nearly invisible to us, HD 44594 is almost identical to our own middle-aged yellow dwarf star, the Sun.)

Around the Sun Our Journey Begins | Probably the best way to understand where we are, and what has changed, is to take a tour of the solar system. Here are some basic facts.

Our solar system has, for the moment, nine planets. There are approximately 2,000 asteroids with a diameter of roughly half a mile or more — of which 410 occasionally drift close by Earth. The entire solar system is surrounded by a giant cloud of icy comets numbering, by some estimates, over one trillion. We'll begin at the very heart of our solar system, the Sun.

At one time we worshiped the Sun, sacrificed to it and constructed elaborate temples and pantheons of religious and mythological characters in its image. For millennia, we have understood intuitively the importance of our nearest star — but we hardly grasped the nature of our relationship with it. Now we study it intently. A host of solar observatories and orbiting spacecraft monitor the Sun's activities,

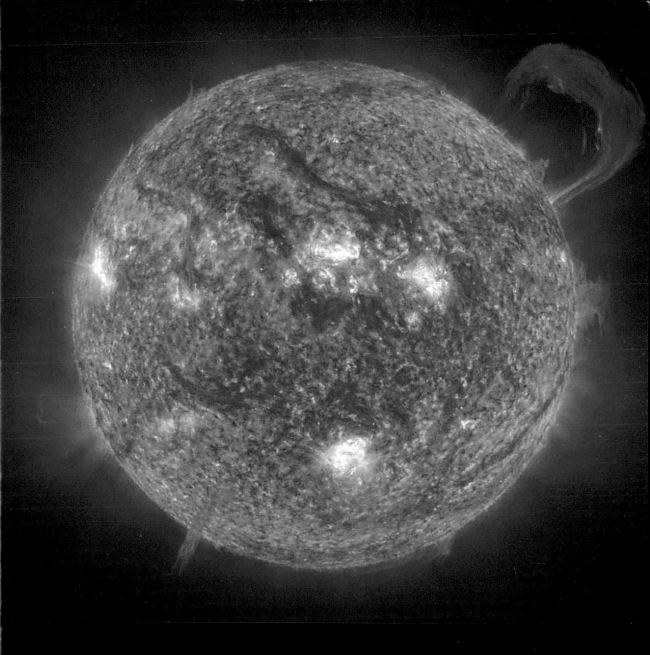

The Sun, with all those planets revolving around it and dependent upon it, can still ripen a bunch of grapes as if it had nothing else in the Universe to do.

— Galileo Galilei

(Left) A radiant corona surrounds the Sun.

(Previous page) This X-ray image from the *SOHO* (*Solar and Heliospheric Observatory*) satellite gives us a dramatic view of our Sun and the knotted, twisted magnetic field surrounding it.

keeping an eye on its eleven-year sunspot cycle and its temperature fluctuations.

Surrounding the disk of the Sun is its corona — those gossamer streamers of radiation seen from Earth during an eclipse. Its surface is a roiling, fiery hell, with a temperature of 6,093°C (11,000°F). Great rivers of fire stretch across the solar surface in every direction. Peppered here and there are zones where the temperature is a little cooler. These are called sunspots, and they are often associated with solar flares. Typical flares and solar prominences are so large that the entire Earth could fit inside one. When a flare erupts, a wave of radioactive particles sweeps out as the loop breaks. These herald the launch of coronal mass ejections, better known as solar storms. Such storms are classified into three categories, according to their intensity in X-rays, and are labeled C, M or X. C-class flares are small and have little effect on Earth, but an M-class flare can boost the auroral lights. The most powerful are X-class flares, which are so potent they can cause electromagnetic blackouts worldwide, disrupting power and communications across the entire planet. On November 4, 2003, the largest flare ever measured — an X28-sized flare — erupted. Luckily, it didn't storm into space in Earth's direction.

Inside the Sun, temperatures are estimated to reach 15 million degrees K (27 million degrees F). The Sun is roughly a million times larger than Earth, and is composed mostly of hydrogen, which it burns at 700 million tons per second. As a comparison, when the ill-fated dirigible *Hindenburg* exploded over Lakehurst,

New Jersey, in 1937, it burned 16 tons of hydrogen. The Sun processes the rough equivalent of 44 million *Hindenburg*s every second!

Unlike the *Hindenburg*, burning hydrogen on the Sun produces helium and balance. The helium is the result of the fusion of two hydrogen atoms; the balance comes from the generation of enough heat to overcome the unceasing inward pull of gravity. The day the Sun runs out of hydrogen, it will start dying. Happily, that day is more than five billion years from now.

The Terrestrial Planets — Our Old Neighborhood | Basking in the warm glow of the Sun are the planets that make up the inner solar system: Mercury, Venus, Earth and Mars. These are known as the terrestrial planets — made primarily of silicate rock and metal, and boasting solid surfaces. Our robotic spacecraft exploration of the inner solar system, our immediate "neighborhood," has been in full swing since the early 1960s.

Mercury | Of our three nearest planetary neighbors, little Mercury, the closest to the Sun, is something of an enigma; though Mercury has been radar-mapped from Earth, only half its surface has been visually charted to date. Because Mercury lies so close to the Sun, a scant thirty-five million miles away, even the powerful *Hubble Space Telescope* cannot tell us much about it. (Point *Hubble* at the little planet and the Sun would burn out its delicate instruments.) That will change by the end of the decade, however, when NASA and the Johns Hopkins Applied Physics Lab launch their *Messenger* probe to this innermost world.

There are some oddities about Mercury that we already know. First, a day on Mercury is longer than a year on Mercury. That's because the planet orbits the Sun faster than it can spin. Mercury is also thought to be extraordinarily dense and composed mostly of iron. Perhaps one of the strangest things in the solar system are the ice caps on Mercury. The *Mariner 10* flyby of Mercury failed to spot a polar ice cap, but the radar images made from Earth revealed a reflectivity at its north pole more consistent with ice than with soil or metal. And if it is ice, how could it survive on a world so close to the surface of the Sun? The answer lies in the fact that the night side of Mercury drops to a chilly -148°C (-235°F). Protected from solar heat by the Mercurial night, the planet's thin atmosphere freezes out and creates a temporary polar frost.

Venus — the Oven World | On Venus, however, there is no ice. The third-brightest object in the sky after the Sun and Moon, Venus is sometimes referred to as the morning star or the evening star. Scientists also label it Earth's evil twin. Although it has nearly the same mass, volume and chemical makeup as our planet, it is different in four notable respects. First, Venus spins in a direction opposite to Earth's, an effect astronomers call retrograde rotation. That means the Sun rises in the west and sets in the east on Venus — exactly backward from Earth. This is a testament to the chaos that must have occurred when the solar system was forming.

Secondly, because of a dense, choking atmosphere composed mostly of carbon dioxide, Venus suffers from a superhot, runaway greenhouse effect. Once we believed Venus to be a tropical paradise, but we now know the unfriendly skies over this planet raise the average daily temperature to a roasting 464°C (882°F). That's hot enough to melt lead. The only Earth emissary to land on this sweltering planet was the Soviet Union's *Venera* spacecraft. Thanks to *Venera*, we have rare ground-level glimpses of this broiling world.

The third big difference between Earth and Venus is that Venus has no magnetosphere, which means that it has no magnetic poles. A compass would be useless there.

Finally, there is the lack of plate tectonics on Venus. On Earth, plate tectonics is the drifting of continental landmasses around the surface of the globe — a process first observed by German astronomer and geophysicist Alfred Wegener. In 1912, Wegener noticed the uncanny way in which the South American coastline mirrored that of Africa. We do not yet know why some worlds have tectonic motion while others don't. On Venus there are some buckling features

NASA's *Magellan Radar Mapper* gave researchers their first global view of the hidden terrain sweltering under Venus's skies. The orange color scheme is a false color derived from the Russian *Venera* landers that touched down on the planet's gravel surface. At the south pole several swatches of solid color show still-uncharted regions that *Magellan* couldn't cover.

found along the ridges and valleys of the Maxwell Montes mountain range, for example, and Fortuna Tessera — but none of the grand rifts and convergence zones that we find on Earth. So the mountain-building processes that created these features on Venus are different from those of Earth. Such churning of the crust is also thought to be an important condition for the rise of life.

Venus is a world in flux. The erosion processes that help shape its landscapes are similar to those of Earth. And that erosion is not limited to features on the ground. The thick atmosphere of Venus is also being eroded — by the solar wind from the Sun.

From 1989 through 1994, NASA's *Magellan Radar Mapper* orbited Venus and gave scientists a set of global radar maps of the roasting Venusian surface. It was from this data that we learned that Venus has neither tectonic-plate motions nor any features associated on Earth with continental drift. We also now know that it has relatively few impact craters (only 843 at last count).

Using its image-processing prowess, the Jet Propulsion Laboratory produced a series of animated movies based on *Magellan*'s topographical data. The result was an amazing sequence that allows the viewer to tour the Venusian landscape. In the future there will be more robotic probes to study Venus, but no human is ever going to visit that oven.

Earth and the Moon | By far the most studied planet in the solar system is our own Earth. A full inventory of the changes in our understanding of our home in space is beyond the scope of this book, for there is no single scientific, political, philosophical or even religious discipline that is not in some way shaped by knowledge gleaned from our global monitoring activities. A couple of things do bear mention here, however.

First, we have learned in recent times that the overall climate of Earth is warming. There is much discord in the scientific community over the root causes of this. Some say it is the result of the industrial revolution; others claim it is a natural cycle of Earth's climate. The trouble is that the evidence points in both directions. We know that in prehistoric times Earth has been through warmer and cooler periods. We also know that at the dawn of human civilization, our planet was just coming out of an ice age. On the other hand, we have been monitoring the rise of greenhouse gases in our atmosphere and the subsequent erosion of Earth's ozone layer. These are a direct result of human activity and pollution. We may not yet understand the ramifications of all of this, but at least we can perceive the change. Whether or not we humans will survive is another question. The history of life on Earth is replete with mass extinctions.

In the last quarter of the twentieth century, we began to understand the dramatic impact that the cosmos might have had during the course of life's evolution on Earth. Luis and Walter Alvarez found evidence of a massive asteroid impact in the Yucatan Peninsula — an event thought to have caused the mass extinction of dinosaurs sixty-five million years ago at the KT boundary.

This boundary between the rocks of the Cretaceous (K) and the

Tertiary (T) periods is rich in iridium. Earth's crustal rocks don't normally have such abundances of iridium, but it is known that some meteorites do. An asteroid impact could have showered the world with its iridium. Layers of rock can be read as a record of natural history — the deeper the layer, the older it is. Below the KT boundary layer is the Cretaceous, which is full of dinosaur fossils. Above it, in the Tertiary, there are no such fossils. Several other mass extinctions similar to KT have occurred throughout the history of life on Earth, and there is increasing evidence that asteroid or cometary impacts may have wrought each of these.

The findings of Luis and Walter Alvarez have helped bolster the case for punctuated evolution. The late biologist Stephen Jay Gould suggested that evolution is not a gradual change in animal and plant forms over time. Instead, it occurs in sudden periods of rapid change, necessitated by dramatic changes to environment and habitat, followed by long periods of stasis.

In general, though, the conditions on Earth are not too hot, nor too cold. It is ironic, therefore, that our nearest cosmic neighbor should be such an utterly lifeless place. As *Apollo* astronaut Buzz Aldrin once said, "The Moon is a place of magnificent desolation." It is the inspiration of lovers and werewolves, and for the time being the only other world we humans have ever set foot upon.

The prevailing theory for the origins of the Moon also involves a massive impact — but one far, far greater than the dinosaur-killer at the KT boundary. Early in the history of our solar system, a large, Mars-sized body collided with Earth, utterly destroying the crust and upper mantle of our planet. The debris from that impact made planetary rings, in which — as William K. Hartmann has suggested — two small moonlets formed. These two moonlets finally collided, forming the one large Moon we see.

We continue to learn about Earth's consort in space. In the late 1990s, two probes — *Clementine* and *Lunar Prospector* — found and confirmed the presence of small amounts of water ice on the Moon. This may bolster the case for establishing a base on the Moon, an observatory or perhaps even a lunar resort.

When Worlds Collide

The solar system we study today is a collection of worlds moving around the Sun in sublime tranquility. Yet each of these worlds has scars that hint at a different past.

The long-held view of a slow accretion of dusty particles gradually building up nascent planets is being replaced by a chaotic vision of a solar system wrought from extraordinary anarchy and violence. It is now believed that large chunks of matter whirling around the Sun collided and coalesced into larger and larger planetary bodies. A new theory about the formation of our Moon lends support to this turbulent scenario.

According to scientists William K. Hartmann and Donald R. Davis, the primordial Earth was rammed by a Mars-sized planetary body, called an impactor, approximately 4.5 billion years ago. This impactor is believed to have struck our world a glancing blow — one that literally shook loose the Earth's crust and gouged its upper mantle layers. This happened at a time before there was life on Earth, but had there been life, it most certainly would have been exterminated.

At the moment of impact (left), an enormous amount of debris was catapulted into space. Some of it formed rings (as shown in the image at right), while the rest of it fell back to Earth in a shower of meteor impacts. Some of the material in these rings stayed in orbit and coalesced into multiple moonlets. These moonlets eventually merged until they formed the single Moon we see today.

A competing theory suggests that our Moon was captured by Earth's gravity. The obstacle that any theory about the origin of the Moon must surmount is the simple fact that our Moon and Earth are so vastly different from each other. Aside from the Moon's lack of air and water, it also lacks Earth's iron core. Computer models have suggested that the iron core of the impactor would have melted into the Earth.

Whether or not the Moon is the result of a collision or a capture, both theories bear mute testimony to the chaos of the early solar system.

The Red Planet | We may never visit Venus, but one day, humans will visit Mars.

Mars is a rugged, cold world lashed by 100-mile-per-hour winds and great dust storms. Named for the Roman god of war, this rust-colored planet has been a source of wonder and mystery throughout the ages. Could life once have existed here? That question has haunted us since 1877, when Giovanni Schiaparelli released his charts of Martian *"canali."* Schiaparelli himself may not have intended the word *canali* to be translated as "a manmade canal," but at a time when the Suez Canal had just been completed and construction on the Panama Canal was under way, that's exactly what happened. If great civilizations built canals, then Mars must be the home of a truly advanced society. In 1895, Boston aristocrat and amateur astronomer Percival Lowell published his findings and confirmed that the canals were indeed real and the work of a race of beings coping with a dying world.

Such an extraordinary finding was vigorously and immediately challenged by the scientific establishment, and although Lowell's claims were never accepted within academic circles, they fired the public imagination — and spawned some of the greatest works of science fiction, including those of Edgar Rice Burroughs and H.G. Wells.

Irrigation canals spanning the Martian globe may never have existed, but there is strong evidence suggesting that water once ebbed and flowed on the surface — perhaps even in lakes and a small ocean in the northern hemisphere, but certainly in streams and runoffs. Recent photos from the *Mars Odyssey* and *Mars Global Surveyor* orbiters hint that water may still flow from the Martian permafrost, along the flanks of ancient craters and canyons. And if life exists on Mars, it must be near the water — for where there is water, there is life.

Whether or not Mars has supported life in the past, there has been some discussion about making Mars more habitable for human beings in a process called terraforming. Chris McKay at the NASA Ames Research Center has proposed the "greening" of Mars — but the process is neither simple nor quick. It is also fraught with philosophical and ethical pitfalls. It is one thing to accidentally contaminate a world with germs on one's equipment; it is quite another to remake a world. We

The Oceans of Mars

Warm tropical breezes, sand and surf… Are you imagining some idyllic spot in the Caribbean? Well, try Mars! According to Steven Squyres, principal investigator for the Mars Exploration Rover Mission, the site currently being explored by the robot rover *Opportunity* is thought to be one of several ancient shorelines on the planet. Scientists believe this site, named Meridiani Planum, may once have been ideal for life — and optimal for making and preserving fossil impressions.

How might these shorelines have appeared on a map? Thanks to the laser altimeter on board the *Mars Orbiter*, scientists can now make an informed guess, based on the range of altitudes found on the Martian surface. They suspect that the lowlands of Mars may have been flooded at one time but the highlands have probably always been above water. (Top, left) This map indicates where scientists think the beaches of Mars might once have been located. (Top, right) The map of Mars' beaches, rendered onto a globe. (Below) A hypothetical view of a little stream on an Earth-like Mars. It will take a lot more research on the ground to fix the shoreline locations on a global scale, but the search for the perfect beach on the Red Planet is already under way.

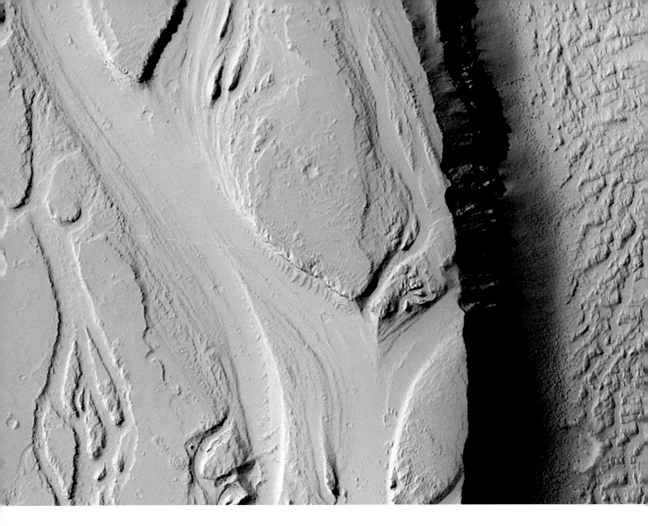

(Above) The Olympica Fossae channels in the northern Tharsis region of Mars, captured by the *Mars Global Surveyor* on October 1, 2003. These canyons and erosion features suggest that water, mud or lava may have flowed here at different times in the planet's history. (Below) Mars during a cyclonic storm. Such events are triggered by a mass of cool air blowing down from the northern arctic region — similar to cold fronts moving across the northern hemisphere of our own planet. Our ability to monitor the weather on Mars provides important insights into the weather here on Earth and helps us understand the climatic systems of a planet we may one day inhabit. (Opposite) Mars as seen by the *Mars Global Surveyor*.

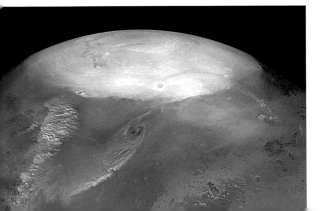

hardly understand the processes by which we're accidentally transforming Earth — in fact, we can't even agree that we are.

In any case, settlers on Mars would have a rugged existence, at least initially. Since 1991, the *Hubble Space Telescope* has been monitoring Martian weather. Recently, it recorded the rise and fall of a global dust storm. The storm began near the Hellas Basin, where it pro- duced titanic walls of Martian dust that churned through the basin and raced across Mars, covering the entire planet for days. A pioneer who lived through such a storm would see the sky obscured by raging clouds of sand and grit.

The exploration of Mars has a checkered history. Since 1960, at least thirty-seven probes have been launched toward Mars, but more than two-thirds have been total failures. The Russians have had a particularly bad run of luck, but the Americans, Japanese and Europeans all have their share of failures as well. Perhaps the most embarrassing was the loss of the *Mars Climate Orbiter* to a metric conversion error. The United States has had some spectacular successes though. *Mariner 4* was the first probe to fly by the Red Planet in 1964. Robotic exploration on the ground began with the two *Viking* landers that touched down on the surface of Mars in 1976, and continued with, first, the *Mars Pathfinder* rover — and, now, the rovers *Spirit* and *Opportunity*. The recent loss of the European *Beagle 2* mission underscores just how difficult getting to Mars has been.

The long run of bad luck in Martian exploration hasn't gone unnoticed. Engineers have begun to chalk up their losses to something they have jokingly dubbed "the Mars Ghoul." In jest or not, this Ghoul has presented a serious obstacle to our survey of the Red Planet.

Roving the Landscape of Mars

When the identical twin robot rovers *Opportunity* and *Spirit* landed on opposite sides of Mars in January 2004, they were expected to operate for about ninety days before succumbing to the harsh Martian environment. (At the time of writing, June 2004, they continue to report dramatic and exciting daily findings to the ground crew on Earth.)

A typical day in the rover's life begins with a wake-up call from its onboard clock. Each rover downloads commands from Earth that detail that day's sequence of tasks. The commands include examining various rocks, operating scientific instruments, taking pictures and driving around. In the afternoon, the rovers use a high-gain antenna to relay collected data to Earth via the *Mars Global Surveyor* and the *Mars Odyssey* spacecraft in orbit around Mars. *Spirit* and *Opportunity* then shut down for the night and await the next command sequence. Meanwhile, the ground crew analyzes the harvest of data and prepares a new list for the following day.

Choosing a landing site for the two Mars rovers wasn't easy. The criteria included low elevations, firm soil so that the rovers wouldn't sink, and a minimum of dust — since substantial dust would block sunlight, rendering useless the solar panels that recharge the rovers' batteries. Although the rover

(Above) An artist's rendering of the Mars rover *Spirit* as it navigates the rock-strewn surface of the planet, gathering samples for analysis.

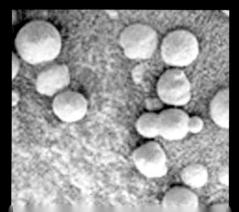

teams wanted places that would maximize the scientific returns, a smooth landing — the most dangerous part of the mission — was also essential. (Inset) *Spirit* touched down in the Gusev Crater, which scientists believe was once filled with water. The lander is visible in the foreground, and the rim of the Bonneville Crater protrudes on the left. *Opportunity* landed on the opposite side of the planet, at the Meridiani Planum, near the planet's equator. (Above) The fresh tracks around the crater — dubbed Fram by ground controllers — were made when *Opportunity* explored the crater's rim. The rover's off-road adventures among craters like Fram also led it to the discovery of numerous tiny spheres that cover the ground (below). These are actually concretions of an iron-laden mineral called hematite, which generally forms only in water. *Opportunity*'s Mossbauer Spectrometer confirmed the presence of hematite, which strongly reinforces the notion that water once stood in this locale. Three spheres that have fused into a "triplet" can be seen in the center of the photo. Such triplets, which are most likely to form only within wet sediments, reinforce our conclusions about the aquatic past of the Meridiani Planum region.

Moving Outward

As we leave Mars behind, we sweep past Phobos, one of the planet's two moons (the other is Deimos). Phobos is an odd, potato-shaped body that may actually be a captured asteroid. This is quite possible, for ahead of us, stretching from Mars all the way to Jupiter, is the Asteroid Belt.

Thought to be the debris field of a planet that never managed to form, thanks to the gravitational pull of Jupiter, the Asteroid Belt is home to more than 700,000 objects (with many more discovered each year). We first learned of asteroids in 1801, when Italian astronomer Giuseppe Piazzi discovered Ceres, but we knew very little about them until recently. Our first close-up view of an asteroid came two days before Halloween in 1991, when the *Galileo* probe bound for Jupiter passed within roughly a thousand miles of the asteroid Gaspra. *Galileo* was able to determine that potato-shaped Gaspra was a mixture of iron, nickel and silicates — typical for asteroids of its type.

Flight controllers got a little surprise in 1994, when *Galileo* passed near another asteroid, Ida. Although Ida is quite small, just thirty-five miles across, the probe discovered that the asteroid

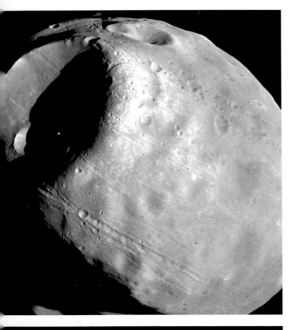

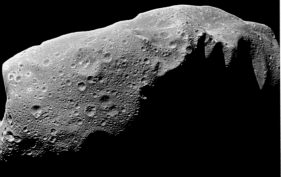

(Top) The Martian moon Phobos, shown in a photo collage from the 1978 *Viking 1 Orbiter* mission. In approximately one hundred million years, Phobos will crash into Mars and kick up enough debris to form a ring.

(Bottom) Asteroid 243 Ida, captured here by the *Galileo* probe during a flyby in 1993, lies adrift in the asteroid belt that stretches between Jupiter and Mars and is thought to be as old as the solar system itself.

boasted its own moon, or moonlet, dubbed Dactyl. (Apparently, researchers rummaged through the annals of Greek mythology hoping to find a character named "Ho," a moniker they thought best suited the satellite of an asteroid named Ida, but to no avail.) As might be expected, Dactyl is tiny — scarcely a mile across.

On February 12, 2001, ground controllers brought the *Near Earth Asteroid Rendezvous* (*NEAR*) spacecraft to a soft landing on the large asteroid 433 Eros, resting it right at the rim of a crater — the amazing finale of an amazing mission. Among other findings, the *NEAR* craft discovered that 433 Eros seemed to have a solid structure, rather than being an accumulation of loose rocks and assorted debris, with about the same density as Earth. Most interesting, the *NEAR* probe photographed a large gouge in the asteroid, perhaps evidence of a collision at some point in its past. Little Dactyl might be the product of just such a collision.

Comets | If asteroids seem placid and stately, comets are the complete opposite.
Unlike asteroids, which can be described as free-floating mountains made of metals and silicates, comets are more akin to huge, rounded icebergs, composed mostly of dirty ices. As a comet approaches the Sun, the side lit by the Sun melts and sublimates, causing the comet to erode. The sunlit side of a comet is in a constant state of upheaval, with enormous geysers bursting through its icy crust. But as the comet's body spins, the exposed side eventually slips back into shadow and refreezes. Meanwhile, the solar wind sweeps the gases and particles jetting off the comet into a long, gossamer tail. Like a windsock fluttering in a stiff breeze, a comet's tail is always pointed away from the Sun.

In 1986, to better understand the nature of comets, various international space agencies launched no fewer than five different probes to rendezvous with Comet Halley. The European Space Agency's *Giotto* made the closest approach, relaying back to Earth some of the most dramatic footage ever seen. *Giotto* showed us Comet Halley as its geysers were blasting out gas, and the probe flew right through the comet's spray.

NASA wasn't involved with any of the Halley probes, but on February 7, 1999,

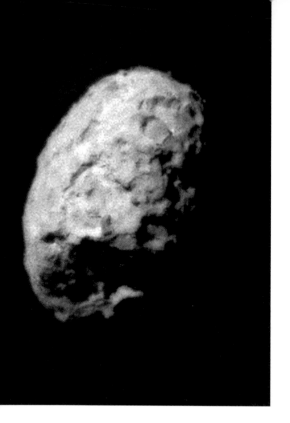

This historic photo of Comet Wild 2 from the *Stardust Sample Return* mission offers the first-ever glimpse of a comet's icy nucleus. Scientists are particularly fascinated by Wild 2 because it is relatively new to the the inner solar system.

to do its own research on comets, its Jet Propulsion Lab launched the *Stardust* spacecraft for a rendezvous with Comet Wild 2 (pronounced Vilt-2). On January 2, 2004, *Stardust* flew within 149 miles of the comet, catching samples of comet particles as it navigated through the particle- and gas-laden haze around Wild 2. It also sent back detailed pictures of the comet's pockmarked surface. The collected particles, stowed in a sample-return capsule on board *Stardust*, will be analysed when the capsule returns to Earth in January 2006.

Comets were once regarded as avatars of bad news. In 1066, a great comet appeared in the skies over Earth, and King Harold lost the Battle of Hastings to William the Conqueror. In 1997, members of the Heaven's Gate cult drank poison as a way to "beam up" to a UFO they believed was hiding behind Comet Hale-Bopp. It is hard to imagine that these balls of ice, swinging around the Sun in their highly elliptical orbits, could have such an ill effect on the affairs of human beings. And yet, the trepidation with which we regard comets may in some respects be well deserved.

In 1908, a mysterious airburst explosion wiped out hundreds of square miles of Siberian forest around the Tunguska River. For three days, unusual orange clouds glowing in the night sky could be seen as far away as Europe. The flattened trees formed a great radial pattern, and it is believed that a small comet exploding in Earth's atmosphere was the cause.

In 1994, *Hubble* recorded some of its most dramatic pictures ever when Jupiter was bombarded by Comet Shoemaker-Levy 9. As the comet headed toward Jupiter, it broke up into a long parade of smaller cometary shards. This parade was dubbed the "string of pearls" by observers on Earth.

As each of these shards slammed into Jupiter's atmosphere, a black mushroom cloud bigger than Earth roiled into the Jovian sky. Meteors and icy comet fragments would have rained across the sky in all directions. Had this happened on our planet, life on Earth would have been nearly destroyed.

Watching Jupiter | Even without comet strikes, Jupiter is a tumultuous world — the largest and closest of the gas-giant planets in our solar system. Gas giants are much larger than Earth, and they have no surface as terrestrial planets do. Jupiter itself is thought to be a failed star, one that never gained enough mass to start the fires of nuclear fusion.

In 1610, Galileo turned his new telescope toward Jupiter and found a miniature system of moons in orbit around the planet, something akin to a miniature solar system. This observation helped bolster the Copernican model of the solar system, since it provided an illustration of the model in action. In Jupiter, Galileo glimpsed a banded, windswept world, distinguished by its Great Red Spot — a cyclone like those on Earth, except that this storm is bigger than three Earths combined. It has also been raging unabated for five hundred years. Unlike Earth's hurricanes, which are low-pressure zones, the Great Red Spot is a high-pressure zone that forms a mound towering over the surrounding cloudscape. Thunderclaps from powerful jolts of lightning slashing across the skies of Jupiter can be heard at radio wavelengths.

One of the most fascinating features of Jupiter is its large, and seemingly ever-growing, collection of moons. At last count, these numbered more than sixty and ranged greatly in size. At the top end, there are the four Galilean moons, so

named because the great astronomer picked them out for the first time in 1610. (Galileo himself named them the Mediciean moons, in honor of the powerful Medici family who helped support his work.) The largest of these moons, Ganymede, has a diameter greater than that of the planet Mercury. In contrast to these, there are many smaller moons, some just a few miles in diameter. In 2003 alone, more than a dozen of these small moonlets were discovered orbiting the gas giant. Since these new moons are appearing faster than we can name them, for the moment they are known only by catalog numbers.

Two hundred and twenty thousand miles above the roiling cloudscapes of Jupiter is the yellow, sulfur-mottled surface of the moon Io. Jupiter is so large that it fills one side of the Ionian sky — a stark contrast to the view of Earth from our own Moon. With lightning storms flickering on its dark side, Jupiter must form a truly dramatic backdrop. Tidal stress caused by the gravitational pull of Jupiter causes Io to be the most volcanically active world in our solar system. When *Voyager I* flew by in March 1979, it photographed one of these volcanoes erupting. Thirty-two years later, the *Galileo* probe navigated straight through one of these plumes. That plume was nearly three hundred miles high and was thought to come from a volcano called Tvashtar near Io's north pole. *Galileo* emerged unscathed, but mission scientists were surprised when they realized the plume came from a new, previously unknown volcano, not Tvashtar.

Europa, another moon of Jupiter, is a place of immense interest to scientists, especially astrobiologists. Riddled with cracks and fissures, it has very few impact craters, meaning it must undergo frequent periods of fracturing, melting and refreezing. Scientists believe that beneath the cracked ice floes of its surface, a vast ocean may ebb and flow under the tidal pull of Jupiter. The same constant squeezing and tugging that keeps Io's volcanoes going must also affect Europa,

Science becomes art in this dramatic image of the moon Io crossing the sunlit side of Jupiter. The clarity of the resolution is a tribute to the power of the *Hubble Space Telescope*. Io is roughly the same size as our own moon and orbits Jupiter at 38,000 MPH. Even though Jupiter's cloud deck is 310,000 miles below, we would expect Jupiter to fill the sky, just as it fills the page here. The dark spot to the right is Io's shadow.

The shattered topography of the Jovian moon Europa is the strongest evidence of a liquid ocean just below the icy surface. The dark lines crisscrossing the surface are fissures, ranging up to 12 miles wide, that most likely are filled with frozen slush.

providing sufficient heat to keep Europa's subsurface waters moving. And where there's liquid water, there is the prospect of life.

Much of what we now know about Io, Europa, and Jupiter and its other moons came to us thanks to the *Galileo* spacecraft launched in October 1989. When it first reached Jupiter in December 1995, *Galileo* parachute-dropped a probe into Jupiter's atmosphere and measured wind speeds over 400 MPH, with gusts topping out at 1,000 MPH. Contrary to what astronomers had predicted, the air was clear, not murky, with visibility for a thousand miles. The probe relayed this data to *Galileo* for ninety minutes before being crushed by overwhelming pressure.

For the next eight years, *Galileo* provided us with a wealth of information about Jupiter and its entourage of moons. On September 21, 2003, scientists at the Jet Propulsion Laboratory sent *Galileo* on a deliberate kamikaze dive into the Jovian atmosphere. They were afraid that when *Galileo* ran out of maneuvering fuel, it might crash into Europa and potentially contaminate it. So, in an act roughly akin

What keeps Europa warm enough to have a liquid ocean? The strong gravitational pull of Jupiter, along with the subtler tug of the other Jovian moons, causes Europa to flex — creating the same tides that occur in oceans on Earth. This flexing keeps hot the upper regions of Europa's mantle (A), which surrounds an iron core (B). The churning mantle in turn keeps the layer of water (C) warm enough to remain liquid. The outermost layer of Europa's water — the layer we can actually see in the photo on page 46 — is frozen (D).

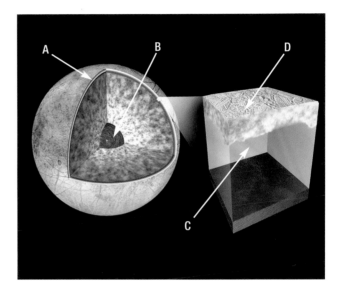

to *Star Trek*'s prime directive (which, in a nutshell, says, "Do not interfere with the natural evolution of alien life"), they chose instead to incinerate the spacecraft — along with any hitchhiking microbes — in the skies above the giant planet.

Galileo has left us with a new riddle concerning Jupiter. The atmospheric probe found large concentrations of argon, krypton and xenon. These gases are known as "noble" gases because they remain aloof and don't easily combine with other elements. (Argon is sometimes used like neon to make dazzling Vegas-style electric signs.) How these gases got there has raised some interesting questions about Jupiter's distant past. Astronomer Tobias Owen of the University of Hawaii has suggested that Jupiter may have formed much farther away from the Sun than where it is today, far out in the Kuiper Belt. The cold temperatures in that far-flung region would make it easier for Jupiter to sweep up those gases.

If indeed Jupiter formed at such a far distance from the Sun and migrated to its present distance, could it be that Jupiter is still slowly drifting toward the Sun?

Saturn and Beyond | Beyond Jupiter lies Saturn. Smaller than Jupiter, its

broad rings make it one of the most distinctive and beautiful planets in the solar system. Galileo described the rings as a pair of orbs flanking the central planet, and then as cup handles. He died not knowing their true nature. It was not until Dutch astronomer Christiaan Huygens developed his theory in 1659 — that Saturn was a world surrounded by rings — that astronomers realized what they were seeing. Huygens believed the rings were solid, but Italian astronomer Giovanni Cassini discovered a gap in the rings that would be named in his honor: the Cassini Gap. The nature of these rings was debated from Galileo's time until well into the eighteenth century, when astronomers concluded that they were composed of dust, ice and rubble. We now know that the rings are made of billions of little ice particles, with traces of silicates and rocks. These particles range in scale from grains of sand to the size of houses, and they form a disk only a few hundred feet thick all the way around Saturn.

As we approach Saturn, we pass the *Cassini-Huygens* probe. At the time of writing, *Cassini* was set to rendezvous with Saturn on July 1, 2004. Along the way, *Cassini* planned to launch its *Huygens* probe toward one of Saturn's moons, Titan. When *Voyager 2* flew by in 1981, evidence of oceans of methane and ethane was found on Titan, but subsequent confirmation has proven difficult because of the thick, opaque atmosphere on that erstwhile world. Scientists believe that the organic chemistry on Titan will provide insights into the organic chemistry of a primordial Earth.

An artist's rendering of *Cassini* as it fires its main engine to allow orbital insertion around Saturn. Main engine burn was to last ninety-six minutes and was viewed as critical to the mission's success.

Saturn as seen by *Cassini* on March 27, 2004. At this point, the probe was still twenty-nine million miles away from Saturn. This was *Cassini*'s last single-image capture of the entire planet and its rings.

Like *Galileo* at Jupiter, *Cassini* will tour the Saturnian system and track the weather on the ringed planet. Although Saturn is normally a quiet world, in 1990 *Hubble* recorded the outbreak of a rare storm on its surface, with wind speeds measured up to 1,100 MPH. This storm spread until it engulfed much of the northern hemisphere of Saturn.

Aside from Mars, Saturn is the planet that most stirs the human imagination, in large part because of its mysterious rings. These planetary rings were once thought to be unique. Then in 1977, astronomers found rings around Uranus, and we now know that Jupiter and Neptune also boast them. But none are as spectacular as Saturn's. Passing through the rings and slipping behind the planet, we can see the Sun as an exceptionally bright, distant star — its light refracted into dazzling hues just as it sets over the Saturnian horizon. The Sun is in fact roughly 840 million miles from Saturn. If we fly across the outer atmosphere of Saturn, we can see that this planet, like Earth, has aurorae flickering in its polar skies.

Saturn now forms a striking silhouette as we leave it behind and head out for Uranus and Neptune. Uranus is unusual in that its rotational pole points toward the Sun, and it rolls around the Sun like a massive bowling ball. What caused Uranus to tilt so far off the ecliptic plane is not certain, but it is suspected that a massive collision with another body may have rocked this world into its current orientation.

When *Voyager 2* passed Uranus in January 1986, it swung quite close to the planet in order to gain a gravity boost that would propel it toward Neptune. Fortuitously, this also brought it by Miranda, one of the moons of Uranus. Little had been known about this moon, but *Voyager's* photographs amazed scientists. Miranda has some of the largest fault canyons in the solar system — crevasses as deep as twelve miles.

In August 1989, *Voyager 2's* journey past Neptune revealed a world far more dynamic than had been anticipated. White-clouded cyclones raced across the surface of this blue-and-aqua-colored world while a Great Dark Spot — another high-pressure zone similar to Jupiter's Great Red Spot — ambled across the southern hemisphere as the planet spun counterclockwise.

We now arrive at Pluto and its moon, Charon. Actually, to call Charon (discovered in 1977) Pluto's moon is a bit of a misnomer. The planet is only about twice as big as its so-called moon, and these brown and gray ice worlds revolve around each other like a pair of spinning nunchucks. Astronomers call this a "double world," but they also debate whether Pluto and Charon are the last planets in our solar system or the first discovered Kuiper Belt Objects. The American Museum of Natural History's Hayden Planetarium in New York City found itself caught in

Charon hovers low in the skies over Pluto. The two bodies are tidally locked, which means that Charon's orbit around Pluto is synchronized with Pluto's rotation. As a result, Charon always appears in the same spot in the sky when viewed from Pluto — never moving across Pluto's skies the way our Moon does across Earth. Recent findings also suggest that Pluto's surface may be rugged and highly varied.

the crossfire when it tried within its exhibit halls to raise public awareness about the debate. The *New York Times* ran the headline, "Pluto Not a Planet? Only in New York." As Hayden Planetarium director Neil Tyson has said, the implication was that somehow far-flung Pluto wasn't big enough to make it in the Big Apple. But the debate continues to this day, with no consensus within the astronomical community.

Recent work by astronomer Marc Buie using the *Hubble Space Telescope* has produced a series of albedo maps, based on the brightness of various features. Until NASA sends a probe to Pluto, these maps provide what may be the best glimpse we will ever have of this remote outpost of our solar system. Meanwhile, NASA has announced plans to send a probe called the *New Horizons* mission to fly past Pluto and explore the realm of the Kuiper Belt. If the mission survives the current round of budget cutting, it should pass Pluto sometime late in the next decade.

As Pluto falls behind us, we are now forty-four astronomical units (AU) from the Sun. That means we're at a distance forty-four times greater than the distance between Earth and the Sun — 4.1 billion miles! It takes light from the Sun a full five and a half hours to reach this lonely, icy world. Radio communications with Earth become a little stilted at this distance: direct a question to Mission Control, and five and half hours later it reaches Earth. The reply comes eleven hours after the question was asked.

Now we plunge into the Kuiper Belt, named for Dutch-American astronomer Gerard Kuiper, who had speculated about its existence in the 1950s. It was not until 1992 that astronomers discovered the first objects that were part of the Belt — leftovers from the creation of the solar system. Most Kuiper Belt Objects (KBOs) are like comet nuclei, mostly snowballs mixed with some rock. Astronomers believe that the Kuiper Belt is the primary source for short-period comets, so called because of the comparatively short time it takes such comets to orbit the Sun. The slightest gravitational nudge from a passing star or planet can start a KBO on an Icarus-like journey toward the Sun.

The New Far Shore of Our Solar System

Since the late 1990s, hundreds of icy bodies have been found lurking along the fringes of the outer solar system — all of them, Kuiper Belt Objects. The latest discovery, announced in March 2004, is the exception. Sedna (pinpointed by arrow in photos, above) is located eight billion miles from the Sun and is thought to be the first Oort Cloud Object. It is so far away that if you were standing on its surface, you could block out the Sun with the head of a pin. Thanks to its extremely elliptical orbit, it takes about 10,500 Earth years to make one year on Sedna. During that year, Sedna's orbit will carry it to a distance of nearly eighty billion miles from the Sun. Right now Sedna is in its closest approach to the Sun, which means that it's summer there, but don't expect any warm breezes. The average surface temperature on Sedna is around -240°C (-400°F).

Catalogued as 2003 VB12, this far-flung body has been unofficially named Sedna, after the Inuit goddess of the sea. It was discovered in a survey of the outer solar system by Michael Brown at the California Institute of Technology in Pasadena, California. Sedna has mystified its discoverers with the slowness of its spin, thought to be the result of gravitational drag from a hypothetical moonlet. However, since there is no other evidence of this moonlet, this is not a widely accepted explanation. Sedna's reddish color is also at odds with the grays and browns exhibited by other ice-bodies such as Varuna, Quaoar, Pluto and Charon. Only Mars is redder in hue.

No one really knows where the Kuiper Belt ends and the Oort Cloud begins — or, for that matter, just how far the Oort Cloud extends. The above illustration suggests that the Kuiper Belt's outer edge blends into the inner regions of the Oort Cloud.

The Kuiper Belt is so far away that direct observation of KBOs is tough because they are dark and small as well as very distant. However, *some* of the larger KBOs have been spotted and even named. Varuna, discovered in November 2000, and Quaoar (pronounced "KWAH-o-ar"), found in October 2002, are each the size of several of America's midwestern states combined. These are the largest KBOs yet discovered (leaving aside the thorny question of Pluto).

As we continue our outbound journey, we see a bright object in the distance. It is a tiny spacecraft, *Pioneer 10*, which is now the farthest manmade object in the cosmos. There are competing claims to the title on behalf of *Voyager 1*, but that's for a spacecraft still running. *Pioneer 10*, whose last, very weak signal was received on January 22, 2003, has been in flight since 1972, cruising at a speed of 27,380 MPH, and is now 79 AU from Earth. Mysteriously, this spacecraft has been speeding up, and scientists don't quite know why — its fuel has long been spent, and its batteries are nearly dead. *Pioneer 10* is headed in the direction of the red star Aldebaran, sixty-eight light years away.

We now find ourselves entering the heliopause, a soft, gossamer solar breeze that surrounds the entire solar system, reaching a distance of 125 AU. This is the realm of our solar "bow shock," where the wind from the Sun encounters the collective winds of other stars in interstellar space. From here, the Sun looks like another star in the sky, and it takes some sixteen hours for its light to reach us.

Finally, we arrive at the Oort Cloud, another enormous repository for dormant comets, reaching out some 50,000 AU from the Sun. Long-period comets, those with very long orbits, begin their sunward trek from these vast realms. The Oort Cloud is a relic of our solar system's formation, and was first suggested by Dutch astronomer Jan Hendrik Oort in 1950 as a way of explaining where long-period comets come from.

Home lies far behind us now. As we head out, we are accompanied by the sounds of yesterday's radio. These broadcasts are the last link with our home in space — the last evidence that we even exist in the cosmos.

CHAPTER THREE

Is There Life in Outer Space?

To consider Earth as the only populated world in
infinite space is as absurd as to assert that in an entire
field sown with millet, only one grain will grow.

— Metrodorus of Chios, fourth century B.C.

As we travel ever deeper into space, we should pause to consider one of the great questions of science: Are we alone in the Universe? For as long as there has been a sentient human mind, there has been the vexing question of just what — or who — lay beyond the horizon. In prehistoric times, our curiosity was pragmatic: food might be found in the next valley, or a hostile neighbor could be preparing a raid. Now we have begun to ponder the greatest horizon of all — the boundless reaches of space. Our search was once for unseen enemies; now it is a hopeful quest for the ultimate Other. It is also an effort to try to understand the meaning of our existence. Are we one civilization within a galactic community? Or are we alone in the Universe?

Until recently, talk of alien life was regarded as the sole province of science fiction and the UFO crowd. The search for "little green men" has required such huge advances in our understanding of so many different arenas — life sciences; geological, environmental and planetary sciences — that any serious undertaking was subject to ridicule and likely to end in quixotic failure. Now those advances have come; consequently, the hunt for extraterrestrial life has not only become a serious topic for science, but has also turned into a frenzied search.

What alien life forms await discovery among the stars? The procession of life as we know it on Earth may help predict what we may find elsewhere in the cosmos.

Gone are the days when political myopia led Senator William Proxmire to bestow his Golden Fleece Award — his way of highlighting and ridiculing particularly harebrained government boondoggles — on SETI (the space program's Search for Extra Terrestrial Intelligence). Astrobiology, the search for life in space, has become the hottest field of astronomical research; so much so, that it is now a central pillar in NASA's exploration of space. Whole branches of science have become reinvigorated, and Nobel laurels await the team that first discovers extraterrestrial microbes or decodes a message from outer space.

Because we've found planets around so many other stars, because we've found such a diverse range of extreme habitats for life here on Earth and because we now have evidence that life arose on Earth so early in its geologic history, scientists now speculate that life may be abundant in the Universe.

A Visitor from Space | In December 1990, Earth was visited by an interplanetary spacecraft. The craft flew to within 35,000 miles of our planet while instruments on board monitored our world for signs of life. During the flyby, the spacecraft detected the presence of ozone in a thin layer in the atmosphere, and a large hole in that layer centered over the South Pole. It found water vapor in the atmosphere, ice at the poles and large bodies of liquid water spanning the globe. These were promising signs, ones that told the interplanetary visitor that our world was suitable for life, since the temperatures were in a range that allowed water to be in liquid form.

The spacecraft's data further showed an unusually high amount of oxygen and methane in the lower atmosphere. This could signify the presence of life, since free oxygen tends to bond with rocks and other gases. Something had to be replenishing that supply of oxygen. Some of the methane could be accounted for by volcanic activity, but not in such abundance as the craft's instruments showed. Gas-producing bacteria, like the kind living in the soil or in the guts of large gas-passing animals, were the likely source. As the spacecraft flew by, it found that regions of Earth not covered by water tended to absorb red and blue light, reflecting back into space only green. This was unusual, since most soil types absorb green and don't reflect it. Some sort of plant life must have arisen. Plant life would also account for the high levels of oxygen.

But while there were signs of microbial life, the probe could not detect any evidence of cities, roads or any other signs of an intelligent, technological civilization. As the spacecraft flew away, however, it did notice modulations in radio waves emanating from the dark side of the planet — modulations showing distinct patterns that natural sources, such as lightning flashes or activity in the planet's magnetosphere, could not account for. No, they had to be signals produced artificially by a technological race of beings.

After finding such a tantalizing clue, the interplanetary spacecraft left Earth behind and proceeded to its next destination, Jupiter. There was no definitive evidence to conclude one way or another that life existed on our world, but there was enough to suggest that closer scrutiny would be justified.

That spacecraft was NASA's very own *Galileo* probe, bound for Jupiter, having just flown in from Venus. Launched in October 1989, *Galileo* used Venus

The *Galileo* spacecraft, shown here sweeping past Jupiter's moon Io. Imagery from the probe has kindled new hope that life might exist elsewhere within our own solar system. Ironically, *Galileo*'s very success inevitably doomed it to destruction. Mission planners, anxious to avoid the possibility of contamination from a crash of the probe on one of the Jovian moons, sent *Galileo* plunging into Jupiter's atmosphere once its work was completed. Their decision remains a controversial one.

first and Earth twice for a gravity boost to gain speed for the long voyage to Jupiter. Carl Sagan suggested using *Galileo*'s flyby of Earth to see if we could detect any signs of life.

Sagan's experiment revealed most strongly the limitations of such robotic flyby missions. Earth, as we know, is teeming with life, so the probe's inability to find definitive evidence, except in the case of the radio modulation, was most startling.

The Search | In our quest to find life in space, what exactly are we looking for?

If we take a cue from science fiction, the world is forever on the verge of being

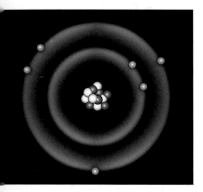

An atom of carbon. All the carbon in the Universe was forged in the furnaces of stars. Take away the carbon, and no living thing would ever exist.

invaded by vastly superior aliens. Orson Welles described such an invasion in a 1938 radio play and set off a national panic. In *Star Trek*, the United Federation of Planets is under constant threat from Klingons, the Borg and myriad other hostile aliens, most of whom are humanoids boasting some sort of latex gill glued to their foreheads.

If we do encounter beings from another world, chances are they won't be quite as silly as this. If we assume that life elsewhere is in a form we would recognize, then what we're looking for is a range of organisms predicated on carbon chemistry in a water medium — organisms just like us.

This is not a foolish, narcissistic assumption. Carbon is found everywhere. The soft graphite in pencil lead is pure carbon. Diamonds, the hardest of all known substances, are another form of pure carbon. Carbon is in the clothes we wear, the gasoline we burn, the plastic we throw away and in every last cell within every living thing.

Because it can accommodate four extra electrons in its outer shell, carbon is capable of bonding with as many as four different kinds of atoms simultaneously. That property makes carbon one of the most versatile elements in the

Universe. When carbon is combined with hydrogen, oxygen, nitrogen and phosphorous, the basic compounds of life itself — the amino acids — can be formed. (Some have speculated that silicon in an ammonia medium could also give rise to a form of life. Silicon does have a respectable range of molecules and molecular chains that it can make. As far as we can tell, however, the rules of chemistry, physics and biology will not permit the natural rise of silicon-based life. It never happened on Earth, and therefore it probably never happened in space.)

The role of water in this carbon chemistry is famous. Life on Earth either lives in water or carries its water around with it: consider the human body, for example, which is seventy percent water. In our quest for extraterrestrial life, we regard water in a liquid form as a prerequisite.

What is it about water that makes it so essential? Water (H_2O) makes an excellent solvent. The oxygen in this important molecule has a slightly positive charge, while the hydrogen has a slightly negative one. As a result, water molecules are able to attach to other atoms and molecules with relative ease. And this allows carbon-bearing organic molecules floating around in water to be exposed easily to lots of other molecules, thereby increasing the number of bonding opportunities. Arguably, guanine, cytosine, adenine and thymine — the nucleotides most important to living things — got their start this way. Sloshing around in water, these complex molecules in turn began to form long chains, turning into macromolecules like ribonucleic acid (RNA) and deoxyribonucleic acid (DNA).

The Structure of Life | For our purposes, we can simply say that life fits into three general categories: single-cell microbes; complex life (like plants and animals); and intelligent beings (language-bearing, tool-making, abstract-thinking but not necessarily technological animals like *Homo sapiens*).

Of these three groups, the microbes are by far the most successful form of life on our planet. Ranging from virophages to single-cell organisms and colonies of single cells like those in stromatolitic mounds, they constitute a very large percentage of all living things.

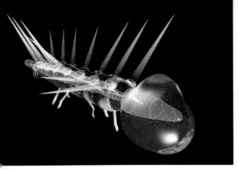

The bizarre form of *Hallucigenia* shown here illustrates an evolutionary dead end on our planet, but a creature similar to this might thrive on another world. If experimentation and change is nature's rule here on Earth, it must be the rule on other life-bearing planets as well.

We've known about microbes for a long time. Antonius van Leeuwenhoek, one of the giants on whose shoulders modern medicine stands, first described in September 1683 the long and short rods of bacteria, the spheres of micrococci and a group of helical-shaped organisms known as spirillum.

On Earth, single-cell organisms come in two flavors: prokaryotic and eukaryotic. Eukaryotes are the kinds of cells that swim around in ponds and make up our bodies. These consist of a nucleus that contains a DNA chromosome, organelles that perform various functions and a casing around the entire package, the cell wall. Prokaryotes are analogous to the nucleus inside the eukaryotic cell — that is, prokaryotes are mostly strands of DNA with a protein casing around them, and that's it. Bacteria and archaea are common forms of prokaryotic microbes.

As far as we can tell, microbial life got a very early start on our planet. Prokaryotic microbes were the first living things on our newly formed world, appearing in the fossil record roughly 3.5 billion years ago. (Some chemical traces have provided controversial evidence that suggests life started even earlier than this, as far back as 3.8 billion years ago.) This transition from organic chemistry to living cells is the greatest evolutionary event in the history of our planet — from dust to living creature in less than 700 million years. The fact that this happened so quickly after the initial coalescing of our planet suggests that in some way, the rise of prokaryotic microbes might just be an automatic byproduct of the planet-making process. Earth's coalescing planetary body, with its heat and water and abundant chemicals, provided a nurturing environment for increasingly complex chemical reactions. As a result, eukaryotic microbes have been around for about two billion years — roughly forty-four percent of Earth's history. By contrast, complex life has been here for only half a billion years — eleven percent of Earth's history.

T'was brillig and the slithy toves, did gyre and gimble in the wabe…

— Lewis Carroll, "Jabberwocky,"

from *Through the Looking Glass and What Alice Found There*, 1872

Where Life Lives | If the rules of nature are the same throughout the Universe, then we can expect the things that caused life here to cause life everywhere — as long as all the conditions are generally the same. Because of the hardy versatility of microbial life, its longevity in the biological history of our planet and the rapidity with which it first appeared, it is likely that if we found another Earth-like world orbiting a star similar to the Sun, whatever its present state of development, we should find microbial life. While simple microbes may occur as a result of planetary creation, the reasons for the rise of complex life are, well, a little more complex. No one is certain why it happened. Cells binding together cooperatively — the essence of complex life — do seem to enjoy an evolutionary advantage, but what triggered this is not known. Under different conditions, it might never have occurred.

This stipulation for finding life similar to that which exists on Earth — that conditions be the same — was long the sticking point for most scientists. Leaving aside the possible existence of Earth-like worlds far out in the galaxy, our neighbors in this solar system were unpromising candidates: too cold, too hot, too dark, too damned dry to support life, given what we knew of it. In the past decade, however, life has been found in the most unexpected places here on Earth.

In the early 1990s, primitive microbes were found living in volcanic vents on the deep-ocean floor. They belong to a class of organisms called hyperthermophiles, and they thrive in extraordinarily hot temperatures, around 399°C (750°F). Their world is dark and sunless, the water boiling, the pressure extreme. Because they live in such a dark realm, their livelihoods do not depend on photosynthesis (the conversion of sunlight into energy) — as do the livelihoods of plants and animals living on the surface — but rather on chemosynthesis (the conversion of chemicals into energy).

Recently, similar primitive microbes have been found living under two miles of solid glacial ice at Lake Vostok in Antarctica. Unlike their deep-sea brethren, the Lake Vostok microbes exist in temperatures far below zero, inside solid ice. Little is known about how they get their food and energy or how they metabolize the little nourishment they do get. Such organisms are oligotrophic; living in a nutrient-poor environment, they cling to surfaces and wait for nutrients to come

Deep-ice core samples from permanently frozen Lake Vostok have revealed a strange menagerie of living organisms. Why such life forms would choose to exist inside ice at the bottom of a glaciated lake is still a great mystery, but scientists believe they provide an excellent analogue for what we might hope to find on Europa and Mars. Radar soundings suggest that Lake Vostok is still liquid at the bottom, and that the ice sheet melts on one side and refreezes on the other — trapping the organisms in the ice when it refreezes.

to them. In the struggle for survival, this gives them one distinct advantage — they don't have to waste any energy, unlike their heterotrophic cousins. Heterotrophic microbes are the more familiar kind that we study in high school biology class. These are the ones you're fighting when you have a cold or wash your hands. They can't produce their own nourishment, so they must expend energy crawling around hunting for food in nutrient-rich environments like the human body.

Some scientists believe that these microbial colonies are but the tip of an iceberg — that a gigantic ecosystem full of untold numbers of microbial species may be found living in the soil for miles into Earth's crust. Until about fifteen years ago, we believed that all living things dwelled upon Earth's surface. Now we believe that nearly half of Earth's entire biomass lives inside the planet's crust.

Life on Earth has such versatility and is so incredibly robust. From extreme heat to extreme cold, darkness and pressure, life's range of habitats extends far beyond the balmy temperatures to which we humans are accustomed. And this raises the prospects of finding life in the extreme environments of other planets within our own solar system. Mars (in particular), Europa and even possibly Venus may also host primitive organisms similar to the prokaryotes found here on Earth.

The Search for Martians | If we ever discover life elsewhere in this solar system, the arena of such a discovery is most likely to be Mars.

In fact, there is already some evidence that microbial life once existed on Mars, even if it no longer does. On August 7, 1996, a bemused White House press corps was treated to a hastily called news conference. President Bill Clinton introduced his science adviser, who then informed the gathered reporters that a fragment of a meteor from Mars, known by its catalog number as Rock ALH 84001, seemed to contain evidence of fossilized microorganisms. The "discovery" remains controversial. (Some wonder whether these so-called microorganisms might not have another explanation — earthly contamination, for example.) Rock ALH 84001

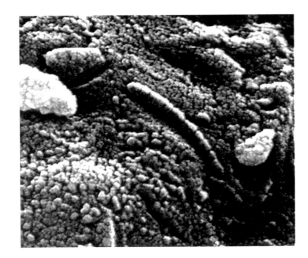

Thought to be a fossil from Mars, the wormy fellow at right is the incarnation of one of the biggest debates in modern science. No strong conclusion can be made either way, but it sure *looks* like a fossil life form. The presence of carbonate minerals and magnetite crystals help bolster arguments supporting researcher David McKay's claim that these are microfossils. Are we seeing something real — or are we simply projecting our hopes onto a random, if freakish, piece of rock? Surprisingly, there is consensus that the rock indeed originated from Mars. Gases trapped inside it are identical to those in the Martian atmosphere.

During the *Apollo 12* Moon mission, astronauts Alan Bean and Pete Conrad retrieved parts of the *Surveyor III* lander for analysis by researchers back on Earth. By then, *Surveyor* had been exposed to the harsh lunar environment for two and a half years. Conrad snapped this photo of Bean standing next to the lander.

aside, there are other compelling reasons to consider Mars. Not least among them is the evidence that water once flowed there. This is clear from the erosion patterns we see on its surface. Recent findings from the *Spirit* and *Opportunity* rovers suggest that a chain of ancient oceans, long since dried up, once existed. It may well be that, under the soil around active geothermal vents, colonies of microbes could still be living.

Our planet Earth has many sites where local conditions are thought to be similar to places on Mars. Many of these analogues have surprised us with the variety and hardiness of the microbes that live there. Mars has virtually all of the same chemicals and materials as Earth and it had for a time a more temperate climate. How can it be, then, that life would have arisen here on Earth and not on Mars? And even if life on Mars were never indigenous, Rock ALH 84001 — and the robustness of microbes presumed living on the *Galileo* probe and found living on the *Surveyor* probe by *Apollo 12* astronauts — would attest to the transportability of life. This planet-hopping capability was totally unknown before the mid-1990s.

Xtreme Fishing | On Europa, there may be no little green men, but what we might find under the surface ice could be as startling as anything contrived in Hollywood.

Europa, an icy moon of Jupiter, is not a friendly place. The average daily temperature is -151°C (-240°F). There is hardly any atmosphere to speak of, and the entire moon is continuously bombarded by such intense radiation from Jupiter that experts believe it would kill an astronaut within minutes. And yet scientists believe that under Europa's icy surface there lies a great ocean, with geothermal vents that could warm and nurture life — just like those on Earth, along the Juan de Fuca Ridge in the Pacific Ocean.

How could Europa have active, undersea volcanism in the first place? Io, a fellow moon of Jupiter, may supply the answer. Io is by far the most volcanically volatile place in our solar system. Like Io, Europa undergoes enormous tidal stress from the gravitational pull of Jupiter and the other moons. This tidal stress actually distorts the shape of the moon into a slight oval. On Io, volcanoes and fissures

> The ice was here, the ice was there, the ice was all around:
> It cracked and growled, and roared and howled, like noises in a swound!
>
> — Samuel Taylor Coleridge, from *The Rime of the Ancient Mariner*, 1797

erupt on the surface as a result. Similar volcanism must occur on Europa as well. The broken rafts of ice we can see on Europa's fractured surface are analogous to what must be occurring on the ocean bed. The surface ice is literally pulled and then squeezed together, forming ridges of immense height and depth.

While Europa's icebound surface blocks out sunlight, it also blocks out Jupiter's harmful radiation — so well, in fact, that some sort of algae could grow just a few feet beneath the surface. In Antarctica, along the Ross Ice Shelf, algae blooms have been found thriving in this most harsh of earthbound habitats. On Europa, it may be possible that the algae would form huge mats, swaying in the ocean currents like vast curtains of kelp and seaweed. Hiding in the mats might be a host of other organisms that either make their homes there or feed off the algae. Perhaps creatures more substantial than basic single-cell organisms seek the warmth around hydrothermal vents. And then, of course, some sort of predator would no doubt have arisen to take advantage of the bounty of Europa's oceans. Think of it — alive, in a cold, dark place along the bottom of uncharted seas, swimming quietly through murky black waters, stalking the hydrothermal vents above the ocean floor. With sudden, lethal precision, it strikes its prey and gobbles down the struggling victim in a single, lightning-fast gulp. In this frigid shadow world of ice and brine, the cycles of life may have turned for billions of years, untroubled

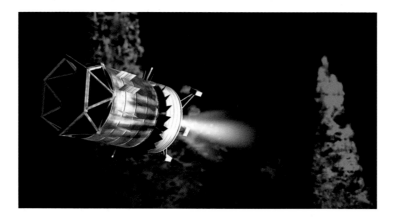

Once it's confirmed that Europa has liquid oceans under its icy surface, a self-sufficient hydrobot — like the one seen here approaching a deep-sea hydrothermal vent — will be dispatched there to search for possible life.

and unobserved. This predator might have developed some sort of luminescence to aid in its hunt. Such an animal could be tiny — or it could be huge. We'll never know unless we go there.

And that's exactly what NASA is planning to do. A reconnaissance mission, dubbed *JIMO* (*Jupiter Icy Moons Orbiter*), will survey Europa, Ganymede and Callisto. After *JIMO*, a future mission to Europa may include a hydrobot — a robotic explorer akin to *Jason*, the minisub that explored the *Titanic*. NASA's hydrobot would have to be one hundred percent autonomous and be able to fend for itself, since no human would be able to recover it. Its mission would be to go fishing, literally, to see whether anything is down there. And if it gets into trouble with some dark, lurking predator under the Europan ice, there will be nothing we can do about it — except celebrate the most profound discovery in human history.

If it turns out that life — even relatively simple microbial life — arose in our solar system in three separate locations, then it must be on planets around other stars. This, above all else, increases the prospects for finding complex life — and, most tantalizingly, intelligent life — elsewhere in the cosmos. Perhaps one branch of life on one of these alien worlds will have evolved into sea creatures that crawled onto the land and planted a flag on their moon. We may find them out in the cosmos, or they may find us.

Intelligent Life | Mr. Jergens had been working late one night at the grocery. As he drove home in his old pickup, he noticed patches of fog wafting through his headlights. Suddenly, just as he was coming around the curve near the Solomons' place, he saw the strangest thing he had ever seen in his life. Three silvery, humanoid figures were standing in the middle of the road.

Jergens pumped his brakes and tapped his high beams. Their silvery bodies gleamed in the headlights as they waddled across the highway. Their arms seemed to be stiff, without elbows, and their legs were without knees. Their heads were tan in appearance, with huge eyes the size of oranges.

And then they were gone. Only the swirling mists were in his headlights. Jergens stopped his truck. That was no deer. He turned around and headed straight for the Highway Patrol office to report an encounter with aliens from another planet.

So, what about intelligent life? Isn't that what we're really looking for? What are the odds of us finding an alien civilization? Or of them finding us?

Again, relying on Earth's history as a means of gauging our chances of discovering life elsewhere in space, we find that the prospects of encountering another intelligent, technological species are pretty low. Not zero, but low. On Earth, the rarest of all life forms has been intelligent life. (When stuck in rush-hour traffic, I sometimes wonder what it will be like when intelligent life *does* arise!) We have only one example — ourselves — in the entire history of our planet. No breed of intelligent dinosaurs, no race of dolphins has ever risen to such a level as humans. It may be the height of *Homo sapien* conceit to point to the obvious, but no other species has harnessed fire or technology or understood the basic processes of the Universe.

More than rare, what we have achieved has been done quite recently. All of our accomplishments and history took place within the geologic wink of an eye — and that's bad news for our search for other alien species. Civilization, the sign by which we can recognize human intelligence, dates no farther back than about eleven thousand years ago. The first Neolithic cities, Catal Huyuk and Jericho, appeared roughly 8,500 years ago. And our greatest technological advances have occurred only in the past hundred or so years: Thomas Edison invented the lightbulb in 1879; Guglielmo Marconi's telegraphy invention, also known as radio, came in 1895; the Wright brothers gave us powered flight in 1903; Enrico Fermi,

Communicating with Aliens

For as long as we have imagined aliens, we have dreamed of communicating with them. And the movies have given us some of our best opening lines. "Gort! Klaatu baraada nikto!" Thanks to a creative Hollywood scriptwriter, Patricia Neal knew exactly what to say to the alien robot when she uttered this now-famous command during the gripping final minutes of Robert Wise's 1951 sci-fi classic, *The Day the Earth Stood Still*. But how do we know which alien "dialect" she was speaking? And to which planets or aliens it applies?

Perhaps aliens are more receptive to conversations set to music, as Steven Spielberg and François Truffaut suggested in *Close Encounters of the Third Kind* — a sequence of five notes repeated in cycles, reminiscent of the best of Philip Glass. Or maybe they prefer numbers. The late astronomer Carl Sagan believed that mathematics might hold the key to understanding extraterrestrial communications — since if we know that two plus two must always equal four, then aliens must know it too.

Let's consider some of the challenges that communication with an alien species presents. All of the above scenarios are predicated on the notion that any exchange would be purposeful and directed. The trouble is, that assumes the aliens already know we're here.

Imagine, for a moment, that you're walking through a dark, thick jungle and suddenly you hear drums. Immediately you wonder: Are they near or far away? Are they friendly or hostile? The drums are obviously a message, but for whom? Now you realize that there may be others crouching in the jungle listening to the drums as well. Should you respond?

The answer to that last question has already been decided. Our radio signals have inadvertently been beating the drums for us ever since Marconi invented radio over a hundred years ago. In fact, our broadcasts have made Earth a brighter source for radio waves than the Sun. Anyone living within an expanding one hundred light-year radius of Earth with a radio receiver powerful enough to pick up our signals knows we're here.

So what happens if, or when, we pick up an alien "drumbeat" in reply? How do we go about deciphering its message?

That's where cryptographers and linguists step in. Thanks to semiotics, we have an analysis of the basic units of language — signs and memes — and they function in both literary and visual media. They work as virtual objects within the symbolic world of language itself. For example, the word "snake" is not a snake but rather the linguistic placeholder of a legless reptile that may or may not be poisonous. Some words may have their origins in onomatopoeia — those words that imitate sounds (the linguistic equivalent of sound effects), such as "buzz" or "ssssssnake." Other words are about grammar and are used in a variety of ways to alter the meaning of the words around them. Both are used to weave a context through different juxtapositions and repetitions.

Visual communication is far more dependent on pictorial onomatopoeia. Much of the history of art is saddled by our desire to mimic reality with a perfect visual representation, but pictures rely upon convention and a small set of grammar rules as well. Whether the first alien message is pictorial or verbal, translating it is going to be daunting — for there will be no shortage of opportunities to get it wrong. For example, art students are trained to see blue as a cool color versus yellow and red as warm colors. For scientists, the opposite is true. And for aliens? The color blue might mean hot, or cold, or something else altogether.

The biggest danger in "translating" any alien communication will be our inclination to project meaning. Humans view language not only as a tool for communication but also — and most importantly — as the very essence of our consciousness. Would it be the same for an alien transmission? Could we get a glimpse for the first time into an alien mind?

Leo Szilard and J. Robert Oppenheimer taught us how to split the atom in 1945; Yuri Gagarin was the first human in space, in 1961; Neil Armstrong, the first human to touch another world, 1969. So much in such a short space of time — and yet the last hundred years represent only 0.000002 percent of Earth's history. Working it out, that means that if we ever find an Earth-like planet out there in space, the odds of it being at a stage of development similar to ours would be five hundred million to one.

Those who have never seen a living Martian can scarcely imagine
the strange horror of its appearance.
— H.G. Wells, from *The War of the Worlds*, 1898

All of this assumes, of course, that the problems that life encountered here would be solved in more or less the same manner elsewhere. Certainly, there are traits that seem to be universal among animals and plants. And yet this is a tricky assumption to make. On Earth, we have multiple cases in which nature has found more than one solution to any given problem. Flying is probably the best example. With wings, nature lifted life off the ground and into the air. But there are so many variations: the thin membranes of dragonfly wings, the feathers of bird wings, the webbing of bat wings. All are completely different designs that successfully solve the same problem. Who can say, then, that on other worlds nature hasn't found even more radical solutions to the same problems life faced here on Earth?

If there *is* anyone else out there, we are not likely to run into them in the flesh. The truth is out there, all right, but it is not what most people believe, or want to hear. UFOs are not from outer space. Whatever they are — clouds, weather, meteors, blotches on film, exotic supersecret warplanes, hallucinations or fantasies — they are not visitors from another world.

At a recent symposium on SETI and extraterrestrial intelligence held at UCLA, Frank Drake — the first person to hunt systematically for extraterrestrial signals, and the inventor of the Drake Equation (a formula for determining the number of inhabited planets) — explained one reason why aliens would not be here: the amount of energy needed to traverse the great distances separating the stars is more than any civilization comparable to ours could produce. A spacecraft traveling from Earth to Alpha Centauri — the nearest star system to our own — within the span of a single human lifetime would need to generate more energy than the entire United States could produce in two hundred years. (Of course, someday, someone might develop a revolutionary technology that allows travel across vast distances. Maybe, somewhere out there, it already exists. But no such technology — that would permit velocities greater than a tenth of the speed of light — is achievable at this point in our history.)

As for the tale of old man Jergens and the three aliens, you might call it a close encounter of a different kind. It happened in the early 1970s, during the uproar over the Pascagoula, Mississippi, UFO abduction (two men fishing from a dock one evening had claimed that they were kidnapped by a flying saucer, triggering a national sensation) and at about the same time as author Erich von Däniken was proposing in his book, *Chariots of the Gods*, that ancient civilizations had been in contact with aliens.

With this backdrop and the recent Moon landings as part of our inspiration, my sister and I had conspired with the kid next door to bring an alien encounter to our town. So we wrapped tin foil around our clothes and attached it with silver duct tape. We put ladies' stockings over our heads and stuck oranges under the nylon over each cheek. Up close we looked as if we had walked off the set of a 1950s B movie, but in the distance, in the fog, at night, we must have looked pretty convincing.

On one level, ours was a little prank, yet on another, deeper level, it was an expression of a yearning to know the truth about life in space. The buffoonery of kids teasing the neighbors. The question is, do we have neighbors?

CHAPTER FOUR

Other Planets, Other Suns

The stars…there's no right or wrong in them.
They're just there.

— Sgt. Elias Grodin (played by Willem Dafoe), *Platoon*, 1986

Our cosmic tour now leads us into interstellar realms. We have left our solar system and are moving into a domain untouched by any aspect of humans or human objects, except for our radio and television broadcasts. We are now in deep space bound for the Alpha Centauri star system, our nearest stellar neighbor. This star system is 4.4 light-years — 26.4 *trillion* miles — away. When we get to Proxima Centauri, the closest star in the Centauri system, the Sun will appear to us as Alpha Centauri appears from Earth — that is, as just another distant star in the sky. The light we see today left these stars 4.4 years ago, a little longer than one term for an American president. This means, of course, that the Alpha Centaurians are only just now getting the returns of the previous election!

As we fly away from the neighborhood of our Sun, we may pass a few stars that, for the moment, have less familiar names: Epsilon Indi and Tau Ceti, both 11.9 light-years from Earth; and 82 Eridani, 19.8 light-years out. These are insignificant, unassuming, yellow dwarf stars much like our Sun. But one of these names may become very important some day if we discover an Earth-like world in orbit there. As we saw in the previous chapter, hope of finding life elsewhere in space now has a broad scientific underpinning, and the discovery of such a planet will most surely raise everyone's expectations.

As we venture beyond the confines of our solar system, questions about life on Mars and Europa lead us to ponder the prospects of finding other worlds

As we move out of our solar system, the view is no longer a familiar one. From the vantage point of a hypothetical planet, we gaze out at a three-star system like Alpha Centauri.

suitable for life in deep space. Astronomers have in fact identified at least six target stars within a twenty-light-year radius of our Sun that may have planets. Scientists will be especially keen if any of these planets is in the "Goldilocks Zone" — a range of distances from the parent star where a planet could be "not too warm" and "not too cold" for liquid water to exist. If such a planet has an atmosphere and an ozone layer, chances are it has life as well. Such a finding would have profound implications, for we would no doubt have to send a reconnaissance probe; on Earth, meanwhile, we might — as Ronald Reagan suggested in 1985 — begin to regard ourselves as citizens not of individual nation-states but of a single planet, Earth.

Eureka! | Science fiction has always assumed planets around other stars but, for most of the history of celestial exploration, we actually knew of only those within our own solar system. Although belief in an infinite number of worlds goes as far back as the ancient Greeks, there was no empirical evidence for their existence. Other planets were at best a fanciful, if unlikely, dream. Only recently has this thinking changed. In late 1995, I got a phone call from *Sky and Telescope* magazine. It seemed that a "stop-the-presses" finding had been made, and the editors wanted to change the cover art for their January 1996 issue. The breathless headline they were planning to run told the whole story: " New Planet Discovered!" How quickly, they wanted to know, could I illustrate a gas-giant planet orbiting very close to its parent star?

Such a discovery now seems commonplace, but in 1995 it was electrifying news. When Michel Mayor and Didier Queloz (both at the University of Geneva) found 51 Pegasus — so called because it is the fifty-first star in the catalogue of

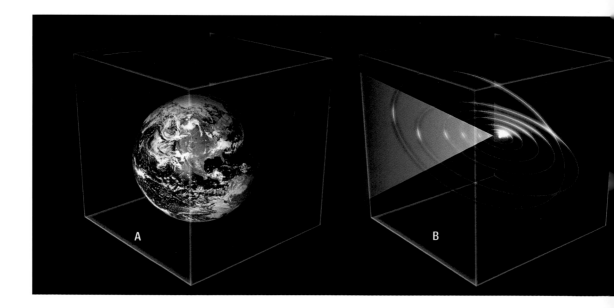

stars within the constellation of Pegasus — they had made one of the greatest discoveries in the history of science and space exploration. Mayor and Queloz have discovered many other planets since — as have Paul Butler and Geoff Marcy at the University of California, Santa Cruz. Thanks to these two groups of planet searchers, we now know (as this book goes to press) of well over 120 new extrasolar gas giants.

The Wobblies | The technique used by Mayor, Queloz, Butler and Marcy to find their new planets is surprisingly simple: they look for stars that wobble. This movement is caused by the gravitational tug of unseen planetary companions.

If we imagine that each stage outward into the cosmos during our "cosmic tour" can be framed inside a box, then each successive box reduces what filled the previous box to "a dot" in a much larger setting. Our journey through the cosmos in this book begins at our home planet (A), moves through the interplanetary space of our solar system in chapter two (B), into the interstellar and pan-galactic space of chapters four and five (C), and finally through the large-scale structure of the entire Universe, described in chapter seven (D). With each step outward, our cosmic perspective changes.

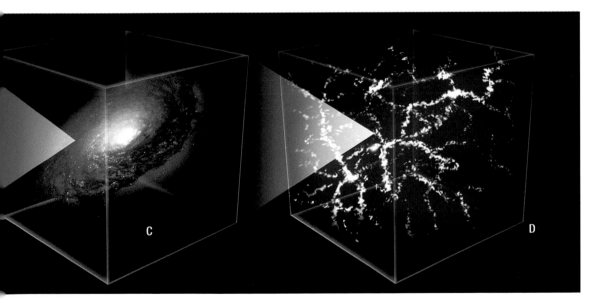

From the strength of the tug and the duration of each wobble, scientists can determine the mass of the planet, the distance from its parent star and how long it takes to make one complete orbit.

Their technique is based on changes in the radial velocity of the star, which is to say that the wobbles we see are Doppler shifts in the starlight. Just as the sound of an approaching train makes a Doppler shift when the train passes, the light from a star shifts from the constant, rhythmic tug of its planetary companion.

This Doppler movement back and forth along the line of sight is good enough to let us identify new solar systems, but it lacks the sensitivity to find smaller, terrestrial worlds like Earth. For that we must wait until NASA launches *Kepler*, a new space-borne telescope named in honor of the great Johannes Kepler, the discoverer of the laws of planetary motion. The telescope will search for transits — the moment when a planet crosses in front of a star. When this happens, the overall brightness of the star dims slightly because the planet blocks some of the light.

The good news is that this is a much more sensitive technique than the Doppler approach, so *Kepler* should be able to find planets similar in mass to Earth (although to do so will be pushing its capabilities). The bad news is that *Kepler* will miss any planets whose orbits are not exactly tilted edge-on to our view from Earth. Unless a planet passes in front of the star — that is, between the star and Earth — *Kepler* will see only the star.

Both of these techniques infer the existence of an orbiting planet from the effect on its parent star. At the Jet Propulsion Lab in Pasadena, California, a radical new telescope capable of making direct observations of Earth-like planets is now on the drafting table. NASA calls this project the *Terrestrial Planet Finder* (*TPF*).

The *TPF* faces some daunting challenges. In fact, making a direct image of a planet orbiting a distant star is so tough that optics experts regard it as one of the greatest possible achievements in their field. Imagine seeing the headlights of a car several miles away and trying to read its license plate. The glare of the headlights must be blocked out, leaving only the reflected light from the plate itself. The solution is a technique called nulling interferometry. This technique combines the light from several telescopes in a way that makes the central star cancel itself out, while allowing photons of light reflected from any potential planets to form into a single image. In this way a group of telescopes works together like one big telescope.

The engineering challenges are formidable. While interferometry is a well-established technique in radio astronomy, optical wavelengths present unique problems. In radio, the wavelengths are much, much longer — as long as buildings are tall and mountains wide. Optical wavelengths, by comparison, range from the size of a needlepoint (red, for example) to the size of the germs living on the head of that needlepoint (in the case of blue).

The delicacy and precision of the finest watchmaker is needed to craft a telescope with that level of pointing accuracy. It is not, as my dad would say, "crude, big-bolt engineering."

Once in operation, *Kepler* and the *TPF* should each reap quite a harvest. Nonetheless, for Jupiter-style gas-giant exoplanets, ground-based observations have us off to a good start. As mentioned earlier, we now count over 120 planets outside our solar system — extrasolar or exoplanets, as astronomers call them — and the number is growing rapidly. In fact, we now know of more planets outside our solar system than within, by a factor of ten. And we have hardly begun to look.

These new exoplanets are generally large (on average, about 3.43 times the mass of Jupiter), but that is likely a statistical bias explained by our current inability to pick out smaller, rocky worlds like our Earth. Historically the distinction between planet and star was seen as clear-cut, but many of these new exoplanets are so large they blur the line. Astronomers generally agree that an object thirteen or more times the mass of Jupiter is no longer a planet. At that size, deuterium — a heavy form of hydrogen that contains a neutron, as well as a single proton, within its atomic nucleus — begins fusing into other elements, qualifying the object as a fully-fledged brown dwarf star. And yet the star HD 202206 has an object orbiting it that boasts a mass 17.5 times that of Jupiter. Still, for all that, it is regarded as a superplanet. The data now strongly suggest that there may in fact be no cutoff between planet and star; instead, there is a continuum from Jupiter to superplanet to planetary mass object to brown dwarf star to red dwarf star.

Our count of brown dwarf stars is also climbing, and this suggests that there may be free-floating planets and superplanets adrift in space, unattached to any particular star. Where these objects came from is not exactly clear; nor, for that matter, is even what to call them. (One sure sign of a robust, expanding science is a debate simply over names. A free-floating planet formed by processes similar to those birthing a star would be called a planetar by one group, a hyperplanet by another, and a sub-brown dwarf by yet another.) Regardless of the appellation, many are unfinished stars that simply never accreted enough mass to start nuclear fusion. An immense collapsing gas cloud, with its massive size and scale, could spawn a large number of such protostars. Perhaps their more massive siblings gravitationally cast them adrift. Another theory suggests that the early stages of a protoplanetary disk are so chaotic that some planets get kicked out after passing too close to a more massive planetary sibling.

The Strange Case of Pulsar B1620–26 | We not only are discovering new planets but we also are learning that some of them have been around for a very long time, far longer than was previously believed possible. In 2003, researchers using the *Hubble Space Telescope* found one of the weirdest planets known to modern science. Everything about this planet is wrong — it is in the wrong place, around the wrong star and it formed at the wrong time.

Located in the core of a globular star cluster called M4, this newfound planet orbits a binary star system consisting of one yellow Sun-like star and a special kind of dead neutron star — a millisecond pulsar — that spins up to 760 times per second (or roughly twenty percent of the speed of light). Globular star clusters are beehives of stars that orbit a parent galaxy. Each cluster consists of tens and hundreds of thousands of stars bound tightly together. The stars themselves are thought to be very old, primarily because they are all low in metalicity (iron and other elements heavier than helium created by fusion reactions inside stars) and other heavy elements. This means globular star clusters should not have planets inside them. Computer simulations tend to support this as well, and even *Hubble*'s own observations of 47 Tucana — a similar cluster orbiting our own Milky Way — showed no planets among its fifty thousand stars.

The M4 planet is approximately thirteen billion years old, which indicates that it was formed when the Universe was very young. But this raises again the issue of metalicity. Heavy elements are thought to have been scarce in the early Universe, so this odd planet proves either that heavy elements were in greater abundance than previously believed — or that the planet-making processes are so efficient that they can work even with a paucity of metals and heavy elements.

Scientists believe that the pulsar captured the planet and its parent star when they drifted into the core of the globular star cluster. That was bad news for this odd planet, for it meant that it would now spend billions of years bombarded by the pulsar's radiation, which sweeps over it like a searchlight a hundred times a second.

The planet of Pulsar B1620–26 is a gas giant like our own Jupiter, though two and a half times as massive. If a Jupiter-like planet can be found so early in the history of our Universe, could the number of Earth-like worlds be greater than we think?

Pillars of Creation and Star Factories | Where do all the stars come from? When the 1990s began, there was a consensus that starbirth began when an older, previous-generation star exploded, sending a shock wave into nearby interstellar clouds. We now know this to be only one scenario. By the end of the decade, significant new progress had been made in understanding the different ways in which stars are born.

The Birth of an Icon | Perhaps the most iconic image of modern astronomy (see p. 82) is Paul Scowen and Jeff Hester's photo of M16, the Eagle Nebula, released in November 1995 from the *Hubble Space Telescope*'s Wide Field and Planetary Camera 2. Part of a star-birthing region 6,500 light-years away in the constellation Serpens, M16 boasts three "pillars of creation" — columns of cool gas and dust.

(Right) A wide-field image of the Eagle Nebula, taken from the Kitt Peak National Observatory. (Above) A detail of the center of the nebula, captured by the *Hubble Space Telescope*. Three "pillars of creation" — columns of cool gas and dust — loom amid a swirl of molecular clouds.

The *Hubble* close-up allows us to see individual stars in the process of being born. The Kitt Peak panorama gives us a sense of the immense scale of this star-forming region. The colors in both pictures help astronomers understand the chemical makeup of the entire region. Both pictures were imaged at wavelengths that feature hydrogen in green, sulfur in red and ionized oxygen as blue. The Kitt Peak image is at similar wavelengths, but the colors are balanced differently.

The tallest pillar on the left measures roughly four light-years from the crest at the top to the bottom edge of the photograph. That's roughly the same distance that lies between our Sun and the star nearest to us in space, Proxima Centauri.

The pillars form when dark and cold molecular clouds are blasted by ultraviolet radiation and strong solar winds from newborn stars. Denser regions erode more slowly than the surrounding cloud, leaving columns like these in place.

The large pink and white star between the leftmost and central pillars may be one of the culprits hollowing out the left-hand pillar. At the top of this feature, there are tiny knots of gas and dust. These are stars in the making. Each tiny clump is roughly the same size as the orbit of Pluto — that is, about 1.4 billion miles across. Jeff Hester has called these clumps EGGs (Evaporating Gaseous Globules), and they represent one of the first stages of star formation.

A similar object representing a slightly later stage in star formation was found in the Orion Nebula by Robert O'Dell, then at Rice University. O'Dell was able to identify several stars not yet born but warming up, each surrounded by a pancake-shaped cocoon of dark gas and dust. Like the clumps in the Eagle Nebula's pillars of creation, these cocoons are roughly the same size as the orbit of Pluto. These were protoplanetary disks, but O'Dell contracted the term to something a little more exotic, yet easier to say — proplyds.

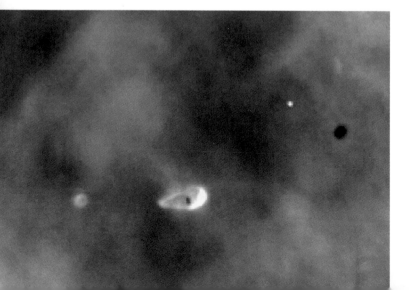

The dark blots (nicknamed proplyds) in this *Hubble* detail of the Orion Nebula are entire new solar systems in the process of being born. At the center of each blot is a reddish pixel, thought to be a nascent star that has not yet undergone nuclear fusion. When fusion begins, each star will shine for billions of years.

A dramatic view of the entire Orion Nebula, taken by astrophotographer David Malin from the Anglo-Australian Observatory in Australia. O'Dell's proplyds (detail shown on opposite page) are approximately one-eighth of an inch below the dark peninsula in the center of the image. This nebula, one of the few that can be seen with the naked eye, can be found as the center star in the Sword of Orion.

O'Dell was very excited. Images of the Orion Nebula had been part of the early release observations from *Hubble* in 1994, and they proved to be a real treasure trove of discovery. The data had just been downloaded from the newly repaired *Hubble* when he and I composited and processed the images. O'Dell was anxious to magnify certain suspicious specks within the image, and sure enough, there were the proplyds — just as he had predicted.

This great nebula can be found in the sword of the constellation of Orion. It looks like a star in the center of a trio of stars running perpendicular to Orion's Belt; but it's really a little star factory, 1.6 light-years across, and some 1,600 light-years away — a nebula that can be seen as a fuzzy patch without the aid of a telescope.

Found in the southern-hemisphere constellation Carinae, some 8,000 light-years away, Eta Carinae is one of the most massive stars in the galaxy and is roughly five million times as bright as our Sun. Astronomers have been watching this pinpoint of light ever since it brightened suddenly in 1837. Over the years, the star's luminosity has fluctuated, and in 1940 its brightness began rising again.

The astronomical community has been divided about exactly what is happening at "Eta Car," but the recent appearance of strontium in the star's spectra has given researchers a warning sign — Eta Carinae is about to blast itself to smithereens. The data suggest that within ten to fifteen years we should witness a catastrophic supernova. And when it blows it will become the third-brightest object in our night sky, behind only the Sun and the Moon. People living in the southern hemisphere of our planet will be subjected to a blast of cosmic rays, mostly neutrinos.

Watching a colossal star like Eta Car in its final death throes has provided astronomers with a windfall of data. Both the *Hubble Space Telescope* and the *Chandra X-ray Observatory* have trained their powerful gazes upon this troubled star. And their images of a catastrophe in the making have yielded both surprises and insights.

Using *Hubble*, we have monitored the growth of two enormous lobes of superheated gases. These lobes are gigantic bubbles, each vastly bigger in size than

(Opposite) The dying star Eta Carinae seen from a nearby orbiting planet. Natural laser light fires out in all directions, created as photons of light stream through clouds of dust and gas surrounding the system.

(Right) *Hubble*'s view of Eta Carinae. Orbiting planets may be pinching the outburst into the double-lobed wasp shape we see here.

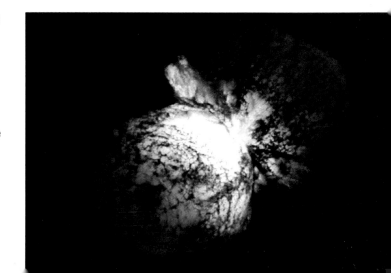

our solar system, swelling at a rate of 1.5 million MPH. *Chandra* found a horseshoe-shaped ring of superheated gas about two light-years in diameter. This ring is evidence of an earlier period of instability that occurred about a thousand years ago.

One of the oddest findings at Eta Carinae are the beams of laser and maser light firing in all directions from near the star's equator. Lasers and masers are rare in nature. It seems that, around Eta Carinae, interstellar gas clouds act as amplification chambers for powerful ultraviolet light rays firing out of the star, focusing them into intense beams.

These strange effects may mark a star near the end of its life. At some point every star must die, and when it does it will take everything with it. Our Sun, the giver of life on Earth, will one day take it all back.

As we travel farther out into our galaxy, we arrive at Beta Pictoris, which many astronomers describe as a primordial disk (a star with a large disk of debris around it). Giant clumps — thought to be protoplanets — can be seen swirling through the disk, with a clearing created by the erosion effects of the star's solar wind.

After Beta Pictoris, we find the Hyades, the closest star cluster to Earth. The Hyades are 630 million years old and were therefore being born at roughly the same time as the rise here on Earth of multicellular plants and animals — the Cambrian Explosion.

As we cruise among the stars, we realize that the space between them is not empty. Ahead lies an ultra-thin wall of hydrogen gas — part of what is called the Local Bubble. This bubble is a relic of at least one ancient supernova that may have caused the formation of many of our neighboring stars, including the Sun.

Now 380 light-years from Earth, we approach a little cluster of stars called the Pleiades. Known to the ancient Greeks as the Seven Sisters, this open star cluster as seen from Earth is shaped like a little dipper and is easily visible in our evening skies. The light we see today left the Pleiades about the same time as Europeans were landing on Plymouth Rock in Massachusetts. Small clouds glowing around the stars are called reflection nebulae. One was photographed by *Hubble* in 2000 and shown to be falling toward the star Merope.

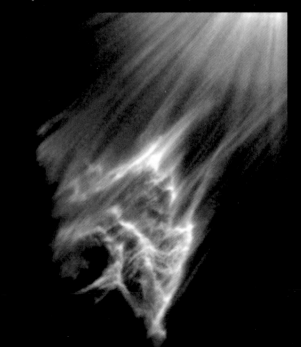

(Above) A portrait of the star Beta Pictoris and the famous dust disk that surrounds it. Like O'Dell's proplyds, this disk was probably formed when the parent star was born. What remains a mystery is why a disk this old never formed planets.

(Inset) *Hubble* sent back this ghostly image of a reflection nebula near the star Merope in the Pleiades, a small open cluster of stars easily spotted in the night sky. This little cloud is falling toward Merope.

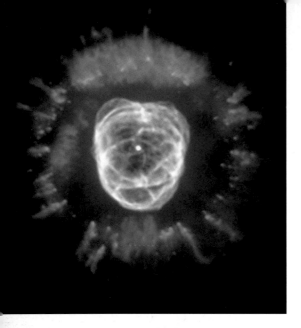

(Left) The Eskimo Nebula, discovered in 1787 by astronomer William Herschel, is so called because our view of it from the ground resembles a person's face in a fur-lined hood. The long filaments in the outer layer are about one light-year long. This dying star offers a glimpse of how our own Sun will one day expire — not in a violent supernova detonation but, rather, in a sudden but relatively calm release of the outer layers of its solar atmosphere.

(Opposite) The *Chandra X-ray Observatory* captured this glimpse of Cassiopeia A, a fifty-million-degree supernova bubble racing into the cosmos. The neutron star in the middle of the bubble had never been seen before.

Five thousand light-years from Earth, we find the glowing remains of a dying star much like our own Sun. Called the Eskimo Nebula, it began forming when this yellow dwarf star in the constellation of Gemini began throwing off its outer atmosphere. This happened at about the same time as the first city on Earth, Jericho, began to rise — nearly 10,000 years ago. We see the bright, dynamic tendrils of gas expanding at wind speeds of more than 900,000 MPH into a double-lobed bubble above and below the star. There may be planets and other debris around this star; if so, they are being incinerated before our very eyes, but they are also slowing the expansion of the gas to a mere 72,000 MPH, giving the nebula its radial, hooded shape.

As beautiful and scary as the death of a Sun-like star is, nothing can compare with the violence and beauty of a supernova — the way stars much more massive than our Sun die. In the Perseus Arm of our Milky Way, thousands upon thousands of light-years from Earth, is Cassiopeia A, a relic of a supernova star death first observed in 1667. Concentric shells of gas surround the burnt core of a once-mighty red giant. This neutron star was first observed by the *Chandra X-ray Observatory* in its inaugural first-light picture in 1999.

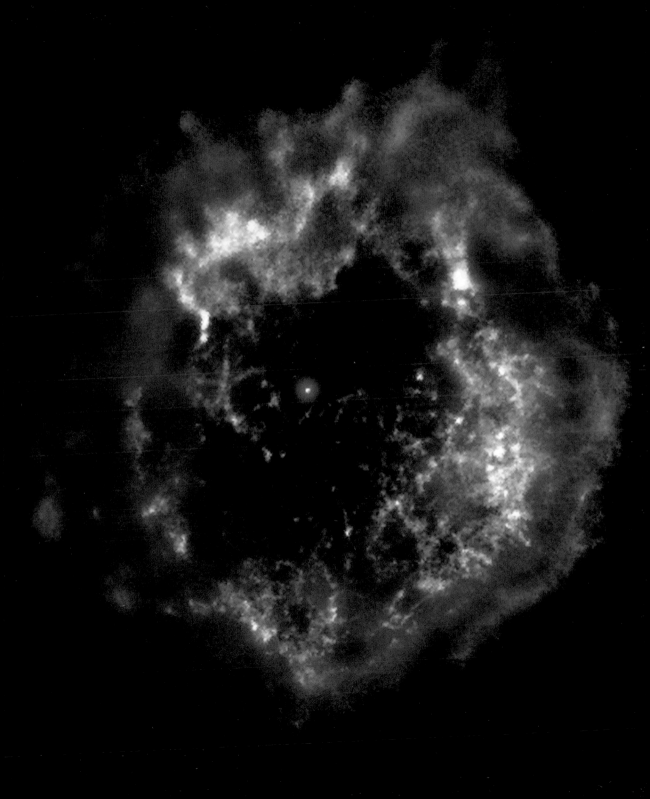

Neutron stars, the cinders of supernova explosions, may find new careers roughing up hapless stars that float within their grasp. Shown here is a millisecond pulsar cannibalizing the Sun-like star at left. The two cones of light are geysers of subatomic particles formed by the powerful magnetic field surrounding the neutron star.

A neutron star is not like our Sun, with its roiling surface; instead, it is a solid body so compacted that even the electron shells of its atoms are crushed, forcing all of the atomic nuclei to combine into one, superdense material called neutronium. If you had one teaspoonful of neutronium, it would weigh millions of tons and would immediately sink to the center of Earth.

A neutron star, with its hard, crystalline surface, can have a pair of ringlike halos above and below an intense magnetic field that surrounds the star. Some neutron stars have been found spinning at extraordinarily fast rates — among the fastest spinning objects known. Some rotate up to 760 times a second, reaching a blistering twenty percent of the speed of light. If the star could spin any faster, it would fly apart. Those that are spinning on their side *and* have an intense magnetic field produce searchlights of intense radiation. Such neutron stars are called pulsars, first discovered with a radio telescope by Jocelyn Bell and Antony Hewish at Cambridge University in 1967. The searchlights have such a perfect beat, it was first thought that they were beacons from "Little Green Men." In fact, Bell and Hewish dubbed their initial discovery, LGM-1, but by the following year, pulsars were found to be neutron stars turned on their sides.

So we progressed, incessantly charmed by some new marvel.

— Jules Verne, *Twenty Thousand Leagues under the Sea*, 1870

To the Core of our Galaxy | As we approach the mysterious realm of the galactic core, we begin to encounter enormous stellar winds, the collective breeze of millions of stars that make up the central bulge. Some intriguing objects lurk here. First is a star cluster called Liller 1 in the constellation Scorpius. Liller 1 is home to the Very Rapid Burster, a binary star system that contains a spinning neutron star.

Surrounding this neutron star is an accretion ring that fuels it. This ring causes outbursts of X-rays, generated both from the material pouring onto the surface of the star and from instabilities within the ring. Each outburst lasts only a few seconds, but the star's rapid spinning causes the outbursts to flicker. In July 2003, scientists used that flickering to determine the star's rate of spin.

Leaving behind the Very Rapid Burster, we now approach a truly strange and enigmatic object — the Great Annihilator. This is thought to be a black hole fifteen times more massive than our Sun. Contrary to the popular belief that black holes only gobble up things, the Great Annihilator is spitting out a torrential flood of energy, so much so that astronomers feel this can only be the result of matter/ antimatter annihilation — hence the name. A thin disk of superheated gases swirls around this black hole, with two geysers of antimatter particles (positrons) racing out along the magnetic poles of the hole. These positrons travel several light-years before they are destroyed by material in the interstellar medium.

Beyond the Great Annihilator is the central hub of the Milky Way. Here lies a supermassive black hole known to astronomers as Sagittarius A West. Roughly 2.6 million times more massive than the Sun, Sag A West makes the Great Annihilator looks like a mere pipsqueak. Its event horizon — that dark, lightless area immediately around the hole — is larger than the orbit of Pluto.

Giant arcs of ionized gas crisscross this monster black hole and form a towering loop, 600 light-years high, roiling above the hole. Scientists believe this is another supernova remnant, perhaps from a star being dragged into the maw of the black hole. This remnant forms a giant bubble called Sagittarius A East.

Surrounding the hole we would expect to find an accretion disk light-years across. On the fringes of this disk we see stars and gas swept in like hapless victims of a great whirlpool. The stars are literally shredded as they fall in. Could this be the engine that drives all of the activity in the Milky Way?

Black Holes: The Great White Sharks of the Cosmos | Black holes
are the most mysterious, fearsome objects in the cosmos. In 1992, an object was photographed in the core of another galaxy — an object thought to be similar to Sagittarius A.

Although NGC 4261 is not a memorable or romantic name, it is one that resonates with astronomers. When the photograph of this object was released from the *Hubble Space Telescope* on November 19, it was hailed as one of the most important images of twentieth-century science.

Captured by *Hubble* is a dark, elliptical shape with a bright dot in the

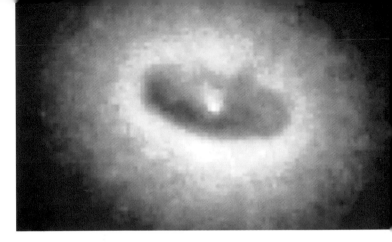

A dust disk, recognizable by its pancake shape, fuels a massive black hole. The disk is found at the core of NGC 4261, a huge elliptical galaxy over 100 milllion light-years away, and may be a relic of a galaxy collision. It is thought to have enough mass to create 100,000 Sun-like stars.

middle. The elliptical shape — a massive swirling disk — spans three hundred light-years and is composed of dust and gases trapped by an extremely compact object over one million times more massive than our Sun. The bright clump in the middle is thought to be the region where this massive compact object resides. In 1992, scientists were hesitant to say what everyone knew intuitively. But Holland Ford, the scientist who made the image, had no other possible explanation: NGC 4261 was a supermassive black hole. His photo remains our closest direct look at this cosmic phenomenon. As Ford put it: "If it isn't a black hole, then I don't know what it is."

Why are black holes one of the most exciting topics in all of science? Perhaps because we are unable to see inside them and have such little grasp of their internal physics, we find them mysterious and menacing. Standing at the top of the cosmological food chain, black holes are to stars what great white sharks are to marine life, or *Tyrannosaurus rex* to dinosaurs.

For the moment, let's consider what a typical black hole actually is. The classic model features a superdense object, called a singularity, surrounded by a zone of blackness called an event horizon. Compacted gases and dust spiraling toward the center usually surround the event horizon, forming an accretion disk. Because of this accretion disk, black holes are not really black; in fact, ironically, they are just the opposite — they're extremely bright.

Matter spiraling toward the hole is compacted and heated up. Friction

(Above) Thanks to observations from the European Space Agency's *XMM* X-ray satellite, astronomers now have evidence of energy being pumped out of a black hole and into the surrounding accretion disk that feeds it. The hundred-million-solar-mass black hole, located in the core of a galaxy known simply by its catalog number MCG 63015, is so powerful that its gravitational pull is dragging the fabric of spacetime around it. According to astronomers Chris Reynolds and Joern Wilms, the magnetic field of the hole is caught up in this drag — and this, in turn, causes the surrounding accretion disk to heat up from friction.

(Inset) Astronomers using *Hubble* were able to capture this image of a jet of subatomic particles racing out from a supermassive black hole at nearly the speed of light. The twisted magnetic field around the black hole pushes streams of charged particles outward along its poles, creating a geyser that shoots across thousands of light-years.

causes the material in the disk to glow in X-rays, and it is this glow, invisible to the naked eye, that telescopes like the *Chandra X-ray Observatory* are hunting for. As well, a black hole sometimes sports a pair of geysers that can be seen shooting in opposite directions along its magnetic poles if the hole is feeding from the disk. Why this happens is not understood. One idea is that some of the material in the disk gets caught in the magnetic field surrounding the hole. This matter traces out the magnetic field and races away along the north and south poles.

Most black holes are the burnt-out remains of dead stars. Stars spend their lives balanced between the inward pull of gravity and the outward pressure of heat. Once a star's fuel has been spent, it can no longer produce enough heat to keep gravity in check — and it suddenly collapses.

The outer layers rush toward the center of the star but, like a storm wave crashing upon a coastal bulwark, the matter rebounds and a shock wave races outward. In one of nature's most beautiful and devastating events, the star goes supernova and explodes. Just as the explosion blows out material in the upper layers of the star, it forces the core to implode. With enough mass, the imploding core will crush under the force of overwhelming gravity. (The classic model is at least three times the mass of our Sun, though recent findings suggest a lower limit may be possible.) If compaction continues unabated, the entire core will be forced into an infinitely small space. Gravity — the force that keeps us on Earth, and Earth bound to the Sun — prevails over all.

It is hard to comprehend just how dense the singularity within a black hole really is. Try to imagine everything in the room you are in — the chair, the furniture, the building, the neighbors, the rest of the town, the continent and the entire planet — squeezed and compacted until it all fits inside the head of a pin. Now crush it down further still so that all of these things can slip inside the nucleus of a single atom. To make a stellar black hole, add into the mix the rest of the planets, the Sun and everything else associated with our solar system. Yet you still don't have enough material to form a black hole typically found in nature. We need the additional mass of at least two more stars like our Sun.

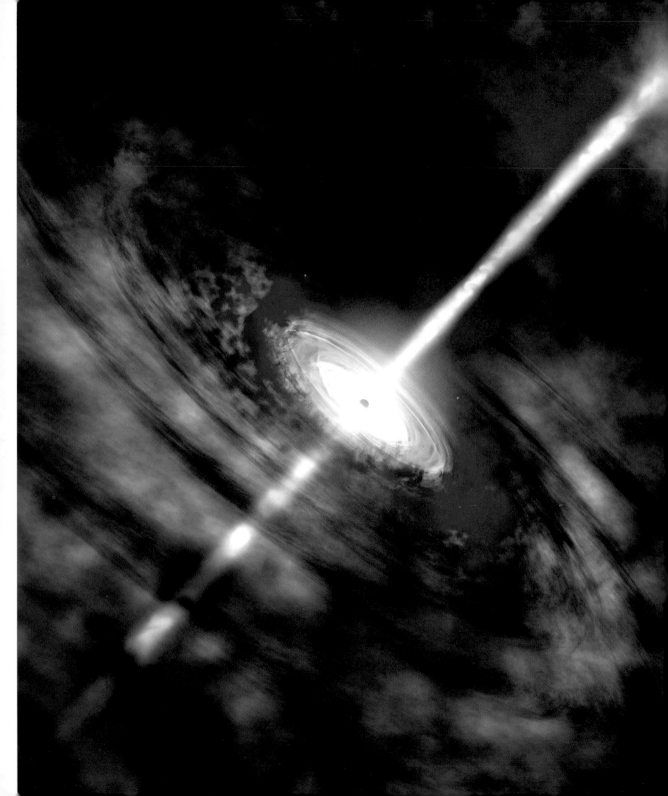

Black Holes:
The Eye of the Storm

The decade of the 1960s was an age of revolution in our comprehension of black holes. During this time, young mavericks such as Stephen Hawking, Kip Thorne and others developed radical theories about the nature of black holes and how they influence their surroundings.

Today we're in a new phase of that revolution. Thanks to technological advances, we are now harvesting empirical data from a variety of probes designed to study the effects of black holes. The exteriors of supermassive black holes are now fairly well understood, for here we have things we can detect and measure with these probes.

The most prominent of these features is a disk of trapped, superheated gases (E) that may include material from extant stars. A small percentage of material in the disk gets snared in a magnetic field surrounding the central region of the disk. When that happens, some of this material is fired outward in a spectacular pair of bipolar jets (D). If the black hole is spinning, there will be a static limit (C). Anything that crosses the static limit will get caught in frame-dragging — the stretching and pulling of space and time around the hole.

The popular conception of a black hole is really the border zone beyond which we have no more information, since light — the conveyor of information — can no longer escape the gravitational pull of the hole. This is called the event horizon (A), which is surrounded by a thermal atmosphere (B) made up of Hawking radiation.

Once material has passed inside the event horizon, it is gone forever. Although what happens inside cannot be observed, we can make a few predictions based on what we know about the initial conditions outside the hole. However, there is a point where the laws of physics break down — at the Cauchy Horizon, which lies somewhere between the event horizon and the center of the hole. Inside the Cauchy Horizon, all forms of predictability are gone. Theory suggests that the singularity — the dimensionless point to which all the mass in the black hole eventually collapses — should form inside the Cauchy Horizon. Is the singularity a knot in space-time or the gateway into another universe? With no physical laws, no observational evidence and stripped of any means of making predictions, we'll never know.

THERMAL ATMOSPHERE

EVENT HORIZON

CAUCHY LIMIT OF PREDICTABILITY

SINGULARITY

When the *Chandra X-ray Observatory* came on-line in 1999, it began turning out X-ray images of black holes all over the sky. *Chandra* found black holes within our own galaxy — and supermassive ones great cosmological distances away. *Chandra* as well as the European Space Agency's *XMM* satellite have shown that black holes existed at a time when the Universe was quite young and were the engines of quasars — those superenergetic objects lighting the earliest epochs of cosmic time.

Small, Medium and Almost Incomprehensible | We have also found

that black holes come in many different sizes. At the top of the scale, there are black holes billions of times more massive than our Sun. The event horizon alone for such a monster is bigger than our entire solar system. Our Milky Way's own Sag A is typical of a class of black holes called supermassive black holes, or SMBHs for short. Objects like these are so huge that they can only be found at the centers of galaxies, and they serve as a foundation around which the rest of the galaxy arranges itself. If an SMBH could suddenly be moved to the outer edge of one of the galactic arms, the rest of the galaxy would re-center itself around the SMBH. In 1974, the British astronomer Sir Martin Rees first proposed that large black holes might exist in the heart of some galaxies. At about the same time, radio astronomers noticed a compact variable radio source near the center of our own galaxy. This radio source resembled the radio images they obtained from quasars — those superbright objects billions of light-years away.

And yet, Sag A is not the biggest SMBH that we know of. In the heart of the giant elliptical galaxy M87, there lurks an object three billion times more massive than our Sun. Such an object is not only large but also old, perhaps even primordial. Black holes of such extraordinary scale are thought to be the engines that drive the activities of a raft of cosmic phenomena — quasars, AGNs (Active Galactic Nuclei) and Seyfert galaxies.

At the bottom of the scale, there is a class of black holes first described by Stephen Hawking as mini black holes. These are thought to be objects with microscopic event horizons and comparatively short lives. They would have formed

This view from the *Chandra X-ray Observatory* of Sagittarius A, the black hole in the center of our own Milky Way galaxy, is from an exposure that took two weeks to make. During this exposure, *Chandra* recorded half a dozen mysterious X-ray flares. Although the cause of these flares is not well understood, we do know they came from the immediate vicinity of the hole.

shortly after the Big Bang, and have probably all evaporated by now. Black hole evaporation, also known as Hawking radiation, is a concept that contradicts our popular notions of black holes and requires quantum physics to explain how anything can escape an event horizon. Virtual particles — one matter, the other antimatter — pop into existence around the threshold of the event horizon. Normally, these particles would touch each other and annihilate, but if they pop in right on the event horizon itself, one of the particles will be drawn into the hole, allowing its companion to escape into space.

As the mass of a black hole goes down, the energy of the Hawking radiation goes up. As a black hole radiates, its mass decreases until it shrinks into its Planck mass, which is roughly equivalent to the mass of a grain of dust. We don't know what happens to a black hole once it reaches its Planck mass.

For a long time, there was a gap in our data between the supermassive black holes and black holes of stellar mass. We expected that the range of black-hole sizes would be continuous, and that there should be black holes with a thousand, ten thousand and a hundred thousand solar masses. Yet we had no examples of medium-sized black holes. Was nature trying to tell us something about black holes, or were we simply missing data?

Then, in 2003, findings from *Hubble* suggested that the star cluster M15 harbors a 4,000-solar-mass black hole, and that the cluster G1 is home to a black hole 20,000 times more massive than our own Sun. These discoveries were the first evidence that we have a full range of black holes.

Looking inside a Black Hole | We cannot actually visit a black hole yet —
and perhaps we never will. Even if we could, we know of no way to survive the trip

Scientists at work on the Large Hadron Collider that will be used to smash particles into an object with the density of a black-hole nucleus.

and report our findings. As the old saying goes, whatever happens in Vegas, stays in Vegas. And anything that falls into a black hole, stays in a black hole. Forever.

Perhaps, however, there is another way to peer inside a black hole. At CERN, the particle physics research center in Geneva, Switzerland, scientists are planning to make an artificial black hole. Using their new Large Hadron Collider, scheduled to come on-line later this decade, scientists will send a particle zipping through the accelerator ring in one direction, and its antiparticle will be fired in the opposite direction. When the two particles collide, for a brief moment their mass and their kinetic energy will be squeezed into the density of a Hawking-style mini black hole. This artificial mini black hole will, for a brief, not-shining moment, let physicists glimpse up close one of nature's most exotic objects without having to travel to the stars.

As with Frankenstein's monster, is there a chance the CERN black hole could escape, ravage the countryside and wreck our planet? Would it remain on the surface of Earth, shredding buildings and all manner of landscapes? Or would it burrow toward the planetary core? How would we stop such a monster, once unleashed? Rest assured, the artificial black hole will be far too small, and too short-lived, to be a threat. And thanks to Hawking radiation, it will evaporate — unless, of course, it finds food.

CERN's ambitious experiment may tell us a lot, but even without it, our thinking about black holes continues to evolve. In 2003, Pawel O. Mazur of the University of South Carolina and Emil Mottola of the Los Alamos National Laboratory proposed a radical new model for black holes. In the classic model, the material in an imploding black hole is completely surrendered to gravity. Mazur and Mottola have proposed that as matter collapses toward the event horizon, gravity reverses itself and becomes a kind of antigravity called dark energy. Dark energy on cosmological scales was first suggested in the late 1990s as a way to explain an accelerating Universe. It gains in strength until, at the border zone of the event horizon, the infalling matter undergoes a quantum phase transition — basically, it freezes out. The event horizon becomes a real surface, forming a shell supported by the force of dark energy. Such an object has been dubbed a gravastar. That gravastars might exist is a tantalizing idea, but for the moment there is no way to test the theory.

Black holes are not the only objects within our galaxy reluctant to yield any secrets. The galaxy as a whole has its riddles, and they may well be intertwined with those of black holes. We are now bound to leave the galaxy and ponder it from afar.

CHAPTER FIVE

The Milky Way and Beyond...

Galaxies are like people. The better you get to know them, the less normal they turn out to be.

— Roger Davies

The Halo Star | On a wall behind astronomer Howard Bond's desk hangs a painting (left) of a greenish, icy landscape filled with hills, cliffs and boulders. This is a quiet, cold, lonely planet in deep space. A gas giant looms in the sky over the horizon, and in the center of the painting is a bright halo star, with a spiral galaxy off to the side. Howard Bond knows this luminous body very well; after all, he discovered it — the single most distant star in our galaxy, the Milky Way.

Imagine for a moment that our galaxy is like a city, full of people and cars, each going about his or her business, to and fro. As we drive from the frenetic, densely populated center of the city toward its outer edges, we pass through sleepy, well-established suburbs. Our Sun is typical of the stars that inhabit these galactic suburbs, roughly two-thirds of the way from the hub of the Milky Way. We drive until we are at the very edge of the city, then we pull in to a rustic gas station on the edge of the desert, the one with the sign that reads, "Last chance for gas." Beyond this point there is nothing but desert for hundreds of miles.

On our tour of the cosmos, Howard Bond's halo star is that final gas station. Look ahead, and there is nothing but vast emptiness. Behind us — just as we would see from that gas station a bustling, sprawling city glittering in the distance — wc now behold the majestic Milky Way, a spiral metropolis of more than two hundred billion stars. Here we are confronted with an extraordinary contrast, one that underscores a fundamental truth of the Universe — that there is something rather than nothing.

Anatomy of a Galaxy | When we find ourselves this far out in the galaxy, we must rethink our sense of scale. Our solar system is measured in astronomical units; 1 AU equals the distance from Earth to the Sun — roughly ninety-three million miles. Pluto is 44 AU from the Sun; and the Oort Cloud extends to 50,000 AU, the far shore of our solar system.

On stellar scales, our entire solar system — all 50,000 AU of it — now appears as a single, dim, yellowish star, easily lost among the myriad stars within our galaxy.

Leaving the solar system to go cruising around the galaxy necessitated our first change in scales; we began measuring distances to the stars within the galaxy by light-years — the distance that a photon of light will travel in a year. That distance is approximately 5.9 trillion miles. Now we add a third, even larger unit, to our measures — the parsec, equaling some 3.26 light-years, or 206,000 AU.

From our lofty vantage point in the galactic halo, we can see all of the major components of a typical spiral galaxy. The most prominent features are the spiral arms — bright areas interrupted by dark lanes of dust and cool gas. These spirals contain billions of stars of every kind at every stage of life and are studded with glowing pink nebulae called H II regions — places where new stars are churned out like exploding fireworks. Together, these disparate arms make up a huge pancake-shaped disk 100,000 light-years across and roughly 1,000 light-years thick.

Gravitational instabilities near the galaxy's core create these spirals and also spawn resonance waves that wash out through the galaxy's disk, piling up gas and dust in the same way that water wave action creates ripples in sand. What

A *Hubble* view of the core of M 51, the Whirlpool Galaxy, reveals breathtaking new detail in a galaxy already known for its photogenic qualities. The Whirlpool is in many respects similar to our own Milky Way. The bright S-shaped band that runs from top to bottom is a pair of spiral arms — and the little pink blotches that stud those arms are massive star-forming regions like the Orion and Keyhole Nebulae. The dark marbling represents lanes of cool gas and dust that provide the raw material for making stars. The brightness comes from the light of a billion suns.

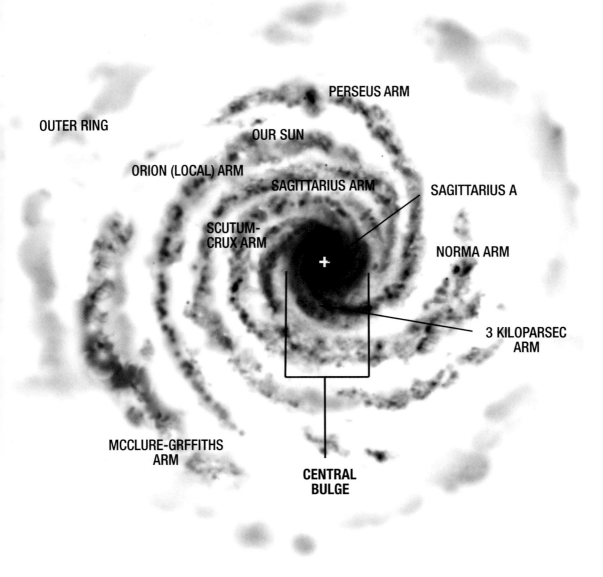

PERSEUS ARM

OUTER RING

OUR SUN

ORION (LOCAL) ARM

SAGITTARIUS ARM

SAGITTARIUS A

SCUTUM-
CRUX ARM

NORMA ARM

+

3 KILOPARSEC
ARM

MCCLURE-GRFFITHS
ARM

CENTRAL
BULGE

Our ability to probe the heavens is allowing us to build, for the first time, fairly accurate charts of the Milky Way.

This panorama view of the core region of the Milky Way, courtesy of the *Chandra X-ray Observatory*, measures 400 by 900 light-years across. Because of obscuring gas and dust, we will never have an optical wavelength view of this region — but X-rays, radio and infrared allow us a peek inside. The bright blob in the center is the supermassive black hole, Sagittarius A. The orange, red and blue fog represents superheated gases several million degrees hot.

we see is the result of this galactic pileup. As gas accumulates, its density increases; stars are born, and they form the spiral patterns.

Recent observations suggest that the Milky Way's disk has a warp in it, quite possibly the result of a recent collision with a dwarf satellite galaxy. Moreover, astronomers recently discovered a mysterious beltway of one hundred million stars surrounding the disk, possibly the product of a galactic fender-bender.

Toward the center of the galaxy we see a huge ball of stars, called the galactic bulge. Roughly 3,300 light-years thick and 20,000 light-years wide, this bulge is populated mostly with older, colder red stars. These stars move through space at different speeds; the closer a star is to the galactic center, the faster it will orbit. The innermost stars are in fact whirling around the galactic core at nine hundred miles per second, roughly one-half of one percent of the speed of light.

Some astronomers contend that the Milky Way has a bar structure within this core region where our supermassive black hole, Sag A, resides.

An intense debate swirls around the origins of galaxies. Do galaxies form from the outside in? Or from the inside out? On the one hand, assuming that supermassive black holes formed outside their host galaxies, it is true that the most massive objects will sink toward the middle of an accumulation of stars, gas and dust. A black hole that got an early start right after decoupling would have had time to draw in sufficient gas and dust to gain supermassive status — with the gravitational strength and appetite to match. Soon enough material is drawn into the vicinity of the hole to form the basis of a galaxy. Could the material that comprises a galaxy be drawn in solely from the strength of a black hole's gravitational pull, or are galaxies formed by pools of dark matter?

Some scientists believe that the cumulative solar winds of stars trapped by SMBHs gain enough strength to turn back infalling gas and dust, thereby retarding the growth of the black hole. Such black holes create clearing zones around them. The immediate neighborhood around the hole is devoid of gas, dust and stars, thanks to the efficiency of the black hole. Any star that ventures close enough to fall in will do so, leaving behind stars in stable, circular orbits that keep them beyond the grasp of the hole. The stars that tend to fall in are those with elliptical orbits around the hole, or whose path is changed by a chance encounter with another star.

The *LISA* mission described in chapter one may help us answer this question. *LISA* will detect gravity waves — those ripples in space-time caused by the sudden shift of extraordinarily massive objects such as stars caught in a death spiral toward black holes. Essentially, *LISA* will hear the screams of stars falling in. It will be up to us to interpret what they mean, especially when the screaming stops.

The Globulars | Hovering above and below the disk of the Milky Way are small beehives of stars called globular star clusters. We know of 138 globulars, each populated with a generation of older stars that may have been born at about the same time as the galaxy itself. We know this because the percentage of heavy metals in a globular's stars is low — in contrast with stars from the same generation as our Sun, which have a comparatively high metalicity.

The origin of blue straggler stars in globular star clusters — those ancient beehives of hundreds of thousands of antiquarian stars, such as M 80 (opposite) — is not well understood, but two theories have emerged. One suggests that the population density near the central region of a globular is so great that some stars strip away the outer layers of other stars. Another idea is that stars in the tumultuous core sometimes collide, and that the new star that emerges is now reinvigorated.

At the heart of one of these clusters, 47 Tucanae, is an unusual star. Although most of the stars in globular clusters are old, 47 Tuc has several stars gently coalescing into one another, forming hot blue straggler stars. As these stragglers take shape, tidal tails are left behind, creating a spiral streamer of glowing red gas — cosmic bunting for a newborn, oddball star. Such mergers are rare, even within the crowded domains of globulars.

Globulars swing in great elliptical orbits around the galactic core, ranging out as far as 100,000 light-years, above and below the galactic disk. Along with loose, unattached stars, they form an enormous halo around the Milky Way.

The components of a galaxy that we've described thus far — the arms, the disk, the black hole in the core, the globulars swinging wide — all have one thing in common: we can see them. But we now know that the visible is only a fraction of what is there. The bulk of our vast galaxy is made up of a mysterious dark matter — and an even more mysterious dark energy. These two entities have profound cosmological consequences, which we will examine in greater depth later. Suffice it to say for the moment that astronomers now believe the Milky Way and its immediate neighbors are embedded within a giant cloud of cold dark matter.

The Local Group | Outbound from the Milky Way, beyond the regions shown

in Howard Bond's painting, our travels bring us to a loose, unstable collection of some thirty galaxies, called the Local Group. Our Milky Way and a second spiral galaxy, Andromeda, are by far the largest galaxies within this grouping. Roughly equal in size and shape to the Milky Way, Andromeda is the only other major galaxy we can see with the naked eye from Earth. All the other galaxies within the group are dwarf satellite galaxies — mostly little ellipticals that resemble fuzz balls in space. Some were discovered recently, like the Sagittarius Dwarf Spheroidal, found in 1994, some 78,000 light-years from Earth.

Andromeda has an interesting distinction in the history of astronomy. For centuries after the invention of the telescope, scientists thought that our galaxy was the total Universe — that the Milky Way was all there was.

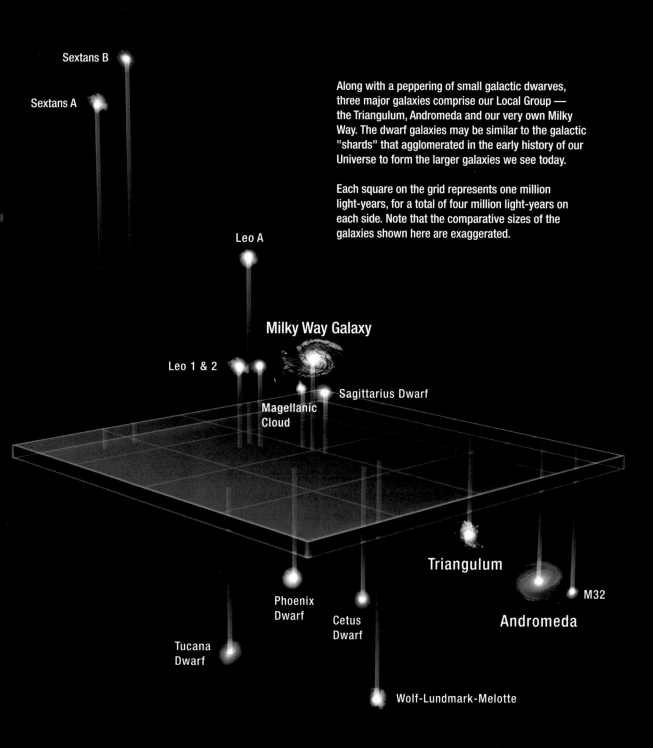

Sextans B

Sextans A

Along with a peppering of small galactic dwarves, three major galaxies comprise our Local Group — the Triangulum, Andromeda and our very own Milky Way. The dwarf galaxies may be similar to the galactic "shards" that agglomerated in the early history of our Universe to form the larger galaxies we see today.

Each square on the grid represents one million light-years, for a total of four million light-years on each side. Note that the comparative sizes of the galaxies shown here are exaggerated.

Leo A

Milky Way Galaxy

Leo 1 & 2

Sagittarius Dwarf

Magellanic
Cloud

Triangulum

M32

Andromeda

Phoenix
Dwarf

Cetus
Dwarf

Tucana
Dwarf

Wolf-Lundmark-Melotte

In 1920 the National Academy of Sciences in Washington, D.C., sponsored a debate entitled The Scale of the Universe. Harlow Shapley and Heber Curtis were invited to promote and defend the prevailing ideas about the nature of spiral nebulae and the Universe at large. The competition between Shapley and Curtis could not have been more acute.

The debate ended in a draw, with each man both gaining and losing something. Shapley argued that the Milky Way was large — 300,000 light-years across. The current estimate is 100,000 light-years. Curtis had argued that the Milky Way was only one-tenth that size.

Shapley also contended that the spiral nebulae were small, star-forming regions, not other galaxies, and that they were all part of the Milky Way. Curtis's opinion was that they *were* other galaxies, as Herschel had suggested, and that they were at great distances.

Then, in 1923, at Mount Wilson Observatory just east of Los Angeles, Edwin Hubble discovered the presence of Cepheid Variable stars in Andromeda. The light from these stars periodically fluctuates in its brightness, allowing scientists to determine their intrinsic brightness. By finding Cepheids in Andromeda as well as in the Milky Way, Hubble proved that Andromeda was another galaxy like our own — thereby validating Curtis's viewpoint. Soon, Cepheids were found in other spiral nebulae, proving that they, too, were other galaxies. Hubble's namesake, the *Hubble Space Telescope*, is still used by astronomers to hunt for Cepheids and to establish distances to the galaxies in which they lie.

The discovery that billions of stars much like our Sun were pooled into great spiral or elliptical structures called galaxies created a sensation. For the first time, people realized the staggering scale of the Universe. It wasn't just the Milky Way; it was made up of countless galaxies *like* the Milky Way.

Resolving the Curtis–Shapley debate would be a satisfying achievement for any career in science, but Edwin Hubble's greatest discovery was yet to be made. While studying the light from these galaxies, Hubble found that they were all redshifted, except for Andromeda, which was blueshifted. Redshift and

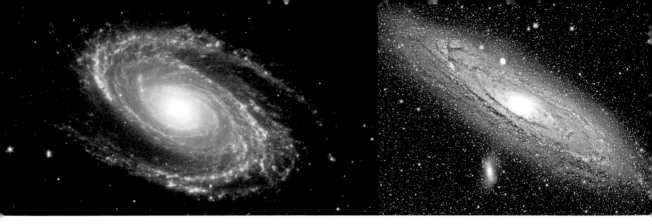

Infrared technology, which registers the heat emissions of, say, a warm human body or the engine in a car, is used on Earth in night-vision goggles and cameras to let us "see" in the dark. Applying that technology in astronomy is the job of the *Spitzer Space Telescope* team (formerly *SIRTF*). In this inaugural image from *Spitzer* (left), we see the heat emission from M81, a typical spiral galaxy 12 million light-years away in the constellation Ursa Major. The warm glow in the arms and the tight knot in the center suggest that there is a lot of activity in those places. M31 (right), better known as Andromeda, is the closest major galaxy to our own Milky Way and can be seen with the naked eye as a fuzzy patch of light. Seeing galaxies like Andromeda in light from different parts of the spectrum offers new insights about these galactic bodies.

blueshift provide a means of determining the direction in which the object emitting the light is going — a bit like the Doppler effect we're exposed to when a fire engine roars by and the tone of the siren suddenly drops. Hubble realized that all the other galaxies — that is, all the other galaxies but one — were receding. The implication was that at one time, long ago, they must all have been joined together in a single, titanic object that exploded. Einstein's theory of relativity predicted such a scenario, but in the days before *Hubble*, Einstein had rejected the idea in favor of a steady-state model. It was Abbé Georges Lemaître, a Belgian priest, who first used Edwin Hubble's data to describe the Big Bang.

And what about Andromeda? Hubble also discovered that it was on a collision course with our own galaxy. The merger with Andromeda will take place in about five billion years, and will take millions of years to unfold. As mentioned previously, scientists believe our galaxy has collided with other smaller galaxies before, and one of Andromeda's companions, G1, appears to be the remnant of a merger with Andromeda.

Chasing Rainbows in an Expanding Universe

How do astronomers use light to divine the secrets of the cosmos? The answer lies in spectroscopy — a fundamental tool of astronomy, and a whole science unto itself. By dividing light into its component colors (like those of a rainbow) and finding the patterns of emission and absorption of visible and non-visible light, scientists can determine the chemical makeup of the light source, its distance and general direction relative to Earth, as well as its density and speed. All of the beautiful imagery that comes out of astronomy is but scientific bunting compared to the information we derive from spectroscopy.

Sir Isaac Newton was the first to discover that light was composed of a spectrum of colors, a full account of which he published in 1704. Red, orange, yellow, green, blue and purple are the primary colors and they correspond to specific wavelengths in the spectrum. The light waves from red are longer than the light waves from blue or purple.

A close look at the spectrum rainbow made with a prism (opposite) shows tiny gaps in that spectrum. These gaps are called Fraunhofer lines and they are the pot of scientific gold that astronomers look for when conducting a spectral analysis. Although Bavarian optics expert Joseph von Fraunhofer discovered the dark lines in 1812, German physicist Gustav Kirchhoff and chemist Robert W. Bunsen discerned their meaning.

Through laboratory experiments, Kirchhoff and Bunsen found that these dark lines had unique patterns that correlated with specific gases. These dark lines were the chemical signatures for each gas and are called absorption lines (A). They also found that burning different materials caused new, bright lines to appear in the spectrum. These are called emission lines (B), and they too yield specific

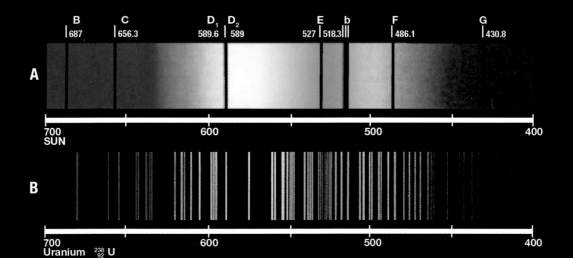

chemical signatures. When Kirchhoff compared the chemical signatures made in the lab with those in the spectrum of the Sun, he found that they were the same, and was able to conclude in 1859 that the Sun was a hot ball of gas. It wasn't long before this technique was applied to other stars.

In the mid-1920s, famed astronomer Edwin Hubble was working on the classification of galaxies when he noticed that many of the chemical signatures for galaxies were shifted into the red — and that the farther away the galaxy, the greater the chemical shift into red. By observing these redshifts in the light wavelengths emitted by the galaxies, Hubble realized that galaxies were moving away from each other at a rate proportional to the distance between them. This finding revolutionized our thinking about the Universe and is now known as Hubble's Law.

In astronomy, the pot of gold is not at the end of the rainbow but *inside* it.

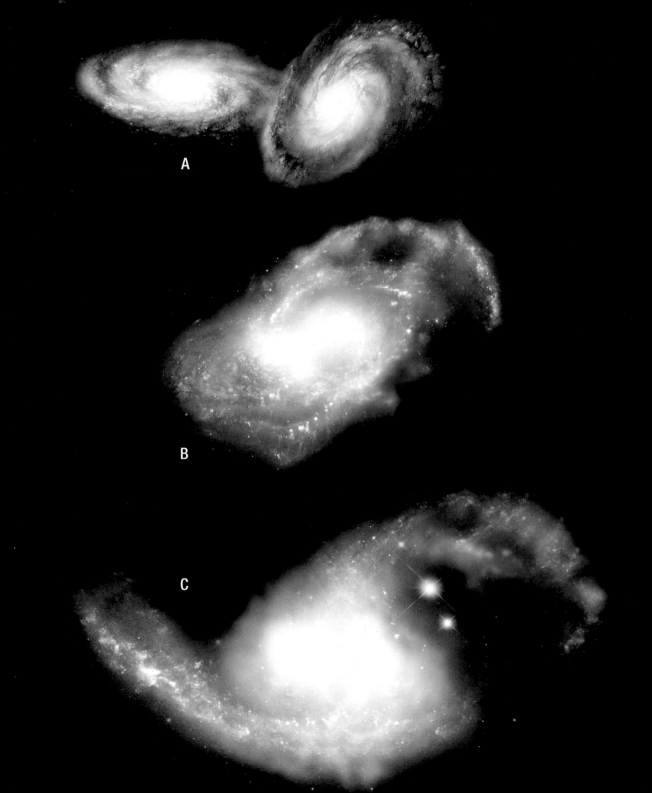

A time-lapse movie of the whole Local Group taken from a sufficient distance will show little galactic pieces coming together to form bigger galactic pieces, as if the entire region were imploding. As the Milky Way and Andromeda galaxies pile up, enormous streams of stars, gas and dust will surge outward, powered by tidal and rotational forces, forming tidal tails. Meanwhile, the core regions will spin around each other in a cosmic do-si-do as new stars form with the rapidity of bursting fireworks. And not just single stars, but whole new star clusters will be born from the cosmic wreckage. The finale will come when the supermassive black holes at the core of each galaxy slam together, launching a series of tsunami-like gravity waves and blasting out a bipolar jet of radiation. The surrounding stars will sort themselves out into a spherical or elliptical halo around the newly merged cores. Finally, for an encore, the other twenty-eight satellite galaxies in the local cluster will also rain down upon the galactic carnage.

One of two possible fates awaits our Earth when the galaxies collide, depending on where Andromeda hits the Milky Way. If our solar system is kicked out of the galaxy in a tidal tail, our Sun and its entourage of planets would be cast into intergalactic space, where they would face a lonely existence. Our evenings would darken, with only a handful of nearby stars to pepper the night sky. The other possibility is that the Sun would get caught up in the melee surrounding the two merging cores. In that case, our evening skies would become far richer. We would see new stars from Andromeda and the infalling satellite galaxies. In addition, in this scenario, the night sky would be studded with hundreds of new stars being born, and we could expect a great increase in the number of stars going supernova — perhaps as many as two or three per year.

A gigantic collision spanning millions of years is set to occur when our Milky Way crashes through the Andromeda galaxy — and there is nothing that can stop it. The time-lapse sequence opposite — based on the work of Matthias Steinmetz (University of Arizona) and Frank Summers (Space Telescope Science Institute) — shows how the merger might play out. The two galaxies fall toward each other (A). During the initial collision, each galaxy's spiral structure is destroyed and tidal arms are flung out from the cosmic wreckage (B). The cores of the two galaxies are drawn together until they cease to exist as separate entities (C) and are blended into a single, homogenized elliptical galaxy. An important consequence of such galactic mergers is that untold numbers of new stars emerge.

Stephan's Quintet comprises five galaxies, four of which are shown in this *Hubble* image. They are undergoing a major period of rapid star formation, called a starburst. Two of the galaxies have already crashed into each other, and a third (cropped at the bottom center of the image) is barreling into the rest.

It is important to note that during this merger, the odds of a direct collision between individual stars are pretty low. The spaces between stars are much greater than the comparative spaces between galaxies. For all of this activity, the skies would remain relatively tranquil, with one exception: stars passing near our system could disturb the Oort Cloud, sending waves of new comets crashing toward Earth.

This description of a galaxy merger is far from fanciful. We have found numerous examples throughout the cosmos. On July 19, 2001, a startling and dramatic image of a famous group of colliding galaxies, called Stephan's Quintet, was released from the *Hubble Space Telescope*. Made famous in the opening of the 1946 film *It's a Wonderful Life*, Stephan's Quintet is 270 million light-years from Earth and has all the features of a classic galaxy merger. This makes the Quintet a fair representation of what we in the Local Group can look forward to in the next three to five billion years.

Hubble's image features a close-up of this tight cluster of galaxies, showing at least two tidal tails and three zones of new star-cluster formation. There is some evidence that several dwarf galaxies associated with the Quintet may have formed out of the galactic wreckage. Many astronomers believe large galaxies are the agglomeration of numerous dwarf galaxies that formed early in the history of the Universe. Analysis of the Quintet suggested that new dwarf galaxies might still be formed as byproducts of galactic mergers — a revelation to many scientists.

A Blizzard of Galaxies | Since Hubble's day, astronomers have carried on finding new galaxies, ever farther out in space. There are more galaxies in the observable Universe than there are individual stars within each galaxy. And they come in different types, too. In addition to spiral galaxies such as ours, there are lenticular, irregular and elliptical ones.

Today we estimate that there are at least 120 billion galaxies that we can observe with our current technologies — and the tally is continuously increasing. That figure is predicated on a seminal series of images NASA released in December 1995 and November 1998 from the *Hubble Space Telescope*'s Wide Field and Planetary Camera 2. The telescope was trained for ten days on an "empty" section of space near the Big Dipper. The Hubble Deep Field (as this, and later the rest of the series, came to be known) revealed more than 1,500 galaxies — all previously unknown. This picture, and the subsequent photos in the series, covered an ordinary patch of sky roughly one-thirtieth the size of the full Moon. If *Hubble* wanted to cover the entire sky in such tiny patches, at ten days per patch, it would take 900,000 years to complete the job! Fortunately, that's not necessary. We can extrapolate.

CHAPTER SIX

The Starry Messengers

We are beholden to give
back to the Universe....
If we make landfall on
another star system, we
become immortal.

— Ray Bradbury, from a speech to
the National School Board Association, 1995

Our journey so far has been a flight of the imagination, steered by the findings of a few scattered robotic probes and the sweep of powerful telescopes across the sky. But as we venture ever farther into the cosmos, we should pause to consider the means needed to explore these distant places firsthand. Robotic exploration of the solar system is well under way, but the current state of manned spaceflight is one of flux, and there is much that we should be doing that we are not. Can humans ever really visit the most distant reaches of our solar system — and do we have the will, the courage and the money needed to do it?

This footprint, made during the *Apollo 11* mission by astronaut Buzz Aldrin, will last longer than the pyramids of Egypt and the cave paintings at Lascaux. With no erosion to wear it away, it should last for millions of years — far, far longer than any other manmade thing. *Apollo 17* astronauts Ronald Evans and Harrison Schmidt left man's last footprints on the Moon in December 1972, in the Taurus-Littrow region, which is roughly two-thirds of the way from the center of a full Moon toward the edge, at about the two o'clock position. Will we one day put new footprints on the Moon, or perhaps on Mars?

The Journey | On December 2, 1993, I had the privilege of seeing the launch of the space shuttle *Endeavour* from Cape Canaveral, Florida.

The stakes for that mission were high — for both the shuttle and the space agency. NASA was sending *Endeavour* on a mission to fix the hobbled *Hubble Space Telescope*. As someone who worked with *Hubble*, I had a vested interest in seeing it fixed.

The shuttle program had had its own share of problems. These ranged from the catastrophic, like the loss of *Challenger* in January 1986, to the just plain annoying, like the time the toilet aboard *Columbia* began to leak. One mission, charged with reeling out a satellite on a 12.5-mile-long tether, had the cable jam after a scant 680 feet had unspooled. But it was the shuttle's difficulty in recovering a spinning Intelsat satellite during a May 1992 flight that was foremost in people's minds.

Working in difficult conditions, shuttle astronauts had captured the wayward spacecraft, strapped on a booster rocket and fired it into the proper orbit. It had been hard work for the crew, requiring additional, unscheduled, time outside the shuttle and a lot of improvisation (Intelsat had never been designed to be serviced in space), but critics were withering about the shuttle's performance. If the shuttle crew couldn't rendezvous with *Hubble* and carry out repairs in space (something the telescope had been especially designed for), then the entire shuttle program might be deemed a failure.

At 4:26 A.M., seven crewmembers strapped inside 125 tons of ceramic tiles, metal and plastic arose on a pillar of flame so bright that the whole sky was instantly bathed in the amber light of an artificial dawn. Once clear of the gantry tower, the shuttle quickly rolled over onto its back and climbed into the firmament.

At approximately one minute after blastoff, the shuttle was little more than a bright star, and the skies were once again dark. A great stalagmite of smoke

The space shuttle *Endeavour* during liftoff. Since it can fly higher and longer than any other craft in the fleet, it was the obvious choice for servicing the *Hubble Space Telescope*. As this book was going to press, NASA was seeking a design to replace the space shuttle.

Two shuttle astronauts execute one of several EVA spacewalks to complete repairs on the *Hubble Space Telescope*, perched in the rear of *Discovery*'s cargo bay during the STS-103 servicing mission in December 1999. *Hubble* was outfitted with six new gyroscopes, batteries, a faster onboard computer and new insulation. Prior to this mission, the orbiting telescope had lost four of its gyroscopes, forcing

snaked into the sky — a mute, vaporous testament to the technological wonder that had just occurred.

Back on the bus, I found myself reflecting on the shuttle's majesty and power. My thoughts turned to a scene from Stanley Kubrick's *2001: A Space Odyssey* in which a protohuman casts a bone into the air and — with the single stroke of ten thousand years, to the strains of the Blue Danube Waltz — the bone turns into an orbiting spacecraft. The shuttle launch embodied the purest essence of all that was good and right about our species, the fulfillment of our dreams and our destiny.

A journey of a thousand miles begins with the first step.

— Chinese proverb

A New Space Race | And yet there is a harsh reality that confronts us now. As impressive as the shuttle was, and continues to be, it is now old. Shuttles have been flying since 1981, and although there have been proposals for a replacement, nothing new has been developed. Originally envisioned as a space truck making frequent trips, the venerable shuttle has proven itself costly, limited and now, with respect to both *Columbia* and *Challenger*, unsafe. The shuttle was meant to simplify space travel, but it wound up being an end unto itself. Whole programs have been forced to accommodate the shuttle's limitations, all for the sake of justifying the huge ongoing expenditures needed to keep it flying. The question is, where do we go from here?

Just a few miles to the south and east of Cape Canaveral are the Bimini Islands, which are part of the Bahamas. Several mysterious shipwrecks in the waters around these islands are now thought to have been part of a great treasure fleet of ocean-going vessels on a mission to circumnavigate the world. The ships were not from Portugal or Spain, or from any European nation. They are believed to be from China, and estimated to have sunk there somewhere between 1421 and 1423 — roughly 70 years before Columbus. The fleet was led by Admiral Zheng He, on behalf of the Ming Dynasty emperor Zhu Di. Under Emperor Zhu, China had

The successful flight of Chinese *taikonaut* Yang Liwei aboard the *Shenzhou-5 (Divine Vessel 5)* heralded the emergence of the first manned space program since the early 1960s. Yang's flight also tested the first new spacecraft design since the Russians abandoned their shuttle-like *Buran* in the 1980s.

a vigorous program of exploration under way, with a massive fleet of over thirteen hundred ships. The largest of the ocean-going vessels were well equipped for long sea voyages, with desalination plants and gardens, and they dwarfed anything that Europe had.

Yet by the time Admiral Zheng returned from his voyage of discovery, China had lost the political will to consolidate its achievements and to exploit its magnificent discoveries. Feeling pressure from enemies both outside and within the empire, the Chinese government mothballed its fleets and destroyed most of the records of Admiral Zheng's epic voyage. China turned away from the world and entered a long period of isolation and technological stagnation.

Today, we may be facing a similar shift. The American space program is without peer in the world, but our manned spaceflight program has lost much of the momentum it had when its focus was beating the Soviet Union to the Moon. American preeminence in space travel is in many ways calcified by a complacency that reflects the modesty of European efforts and the dwindling of the once-mighty Soviet program.

America should pay attention to a new player in the league of spacefaring nations — the People's Republic of China. On October 16, 2003, China's first *taikonaut*, Yang Liwei, went into space aboard an entirely indigenous spacecraft, the *Shenzhou-5*. Loosely modeled after the Russian *Soyuz* space capsule, the Chinese design incorporates some of the lessons learned the hard way by the Americans and the Russians, and relies on the tried-and-true. It presents no great advance in spacecraft engineering, but its no-nonsense simplicity may offer the very reliability the American shuttle lacks.

Could it be that five hundred years after Admiral Zheng and Emperor Zhu, the Chinese have decided to begin a new program of exploration? Recently, they announced an ambitious program for the conquest of space, and so far, cooperation with the West is not part of their plans. Whether they are able to follow through remains to be seen, especially given the expense of human spaceflight. Although little is known of any specific plans for a trip to Mars or perhaps a return to the Moon, the East has nonetheless signaled a willingness to compete for the high ground. A trip to the Moon would be of immense interest here, since today the United States lacks the capability to return there. The prospect that Mars could indeed become the "Red" Planet may seem far-fetched today, but the idea itself may be the impetus we need to trigger a new space race — one to reach that windswept world and expand into the solar system.

While Lieutenant Colonel Yang's debut in low Earth orbit is a positive development, some things haven't changed. The solar system is still as large as it was before. Space is still the most hostile environment there is. And spaceflight is still not cheap. The question is, will America repeat the history of fifteenth-century China?

The notion of looking on at life has always been hateful to me.
What am I if I am not a participant? In order to be, I must participate.

— Antoine de Saint-Exupery

Mars or Bust | The future of manned spaceflight in the short term can be summed up in four words: The International Space Station. The station is an engineering marvel but, like China's *Shenzhou-5* space capsule, it represents no technological leap forward. Nonetheless, it does allow space travelers to rehearse for long-duration

spaceflight, such as to Mars. Like the *Apollo* moon missions, a trip to Mars will be heavily choreographed, allowing as little room as possible for surprises. As the Norwegian polar explorer Roald Amundsen said, "Adventures happen to the incompetent."

Rehearsals of a potential Mars mission have been underway since 1997. On Devon Island, in the Canadian Arctic, a team of hardy explorers has studied what it might be like to live on Mars. At the rim of an ancient meteor crater, these explorers have established an analogue for a Mars base. A white cylinder, the size of a bus, stands on end over a pristine, wind-whipped landscape and represents the habitation module of a Mars Lander. Robert Zubrin, a former Lockheed aerospace engineer, and his Mars Society have based themselves on Devon Island because the weather conditions are similar to those on Mars. Could human beings live and work in such close quarters under extremely harsh conditions?

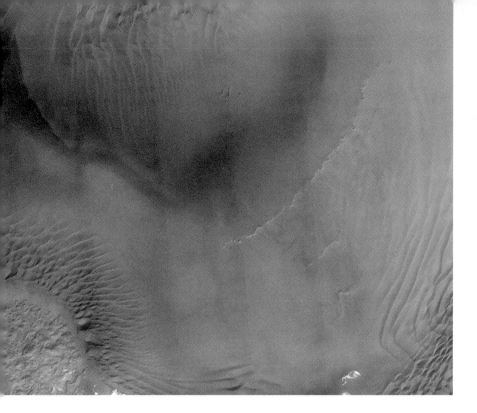

The Juventae Chasma on Mars may prove to be an ideal location to search for fossils. The region has a range of geologic features, including what may once have been beaches, rampart craters and sand dunes.

The lander is part of a larger plan Zubrin has devised for the exploration of the Red Planet. Once the perilous journey to Mars has been safely accomplished and his crew has settled down in the real Mars Lander, one of their primary tasks will be to search for microbial life just under the Martian soil. Contamination will be a critical concern, since both the health of the scientists and the pristine, native condition of the samples must be ensured.

Another task will be to hunt for fossils that may be deposited in riverbeds, volcanic vents and dry chasms. Obvious places to look include the great canyon complex of Valles Marineris, especially around locations like the Juventae Chasma, where multiple layers of sedimentary rock lie exposed. Any rampart craters — craters whose flanks were made by the ejection of soft mud — would also be suitable sites. So far, Zubrin's plan looks promising — so promising, in fact, that many of its key elements have been adapted by NASA in what is known as the Mars

Reference Mission, a hypothetical mission design that serves as a template for future expedition planners. A fundamental aspect of both Zubrin's plan and the Mars Reference Mission is the idea of splitting the mission and, as Zubrin puts it, "living off the land."

This split-mission strategy begins by sending a robotic fuel factory to Mars. The factory extracts all the ingredients it needs to make fuel for the return trip to Earth. Included with the factory is an Earth Return Module. Before any humans blast off from Earth, their ride home will be gassed up and waiting for them on Mars. Such an approach would immensely reduce the cost of the whole Martian venture and help ensure the safety of the explorers.

The flight to Mars will take roughly 130 to 180 days if the planet is at opposition. In the original von Braun vision, a huge Mars-bound spacecraft would be assembled in low Earth orbit with the help of an orbiting space station. Neither Zubrin's plan nor the Mars Reference Mission calls for such a detour. All the components are sent directly to Mars with no stops along the way. Zubrin even calls his plan Mars Direct. The beauty of the split-mission approach is that only the humans need to be worried about the shortest possible transit. The fuel factory, return vehicle and scientific equipment, riding on a separate spacecraft, can take as long as they want to get there. Some of that equipment would include little trucks, an airplane and perhaps one or more helper robots. The Mars Reference Mission calls for a stay of six hundred days on Mars, a little more than a year and a half. Zubrin has made a compelling case for Mars, and perhaps, with a human presence on location, we may begin to answer some fundamental questions about life on the Red Planet.

Once the planet has been thoroughly reconnoitered, we may start the long process of terraforming Mars — that is, re-creating it in the image of Earth. Freeman Dyson has suggested that plants can be engineered for the specific task of Mars terraforming. The plants would become "warm-blooded," and could help transform the Martian atmosphere into something better suited for humans. This is far in the future, however.

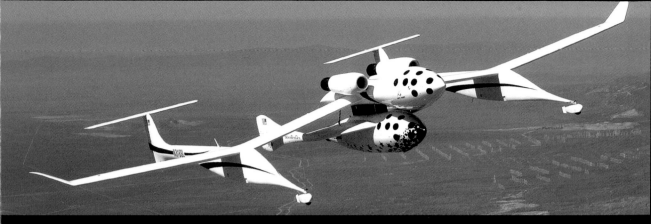

"Wanna take a ride?"

This provocative invitation to a journey into space — issued by the mysterious billionaire Hadden Suit in the 1997 film *Contact* and scripted by Carl Sagan — continues to fire the individual human imagination, as well as the collective.

With a couple of exceptions, spaceflight has so far been restricted to government-funded space programs — and only professional astronauts or military personnel can take part in space missions. The exceptions are California businessman Dennis Tito and South African Mark Shuttleworth, who each paid $20 million to ride the Russian *Soyuz* to the International Space Station (in 2001 and 2002, respectively), much to the consternation of NASA. The willingness of Tito and Shuttleworth to pay such an exorbitant fare underscores a growing desire for nonprofessional, civilian access to space — a desire that has long been neglected.

Until now. A $10-million award, aptly called the X Prize, is being offered by space advocate Peter Diamandis and his X Prize Foundation as a means of jump-starting private, civilian spaceflight. Diamandis has succeeded in rekindling the same spirit of adventure and competition that fuelled the glory days of transatlantic airplane races — including Charles Lindbergh's famous solo nonstop flight from New York to Paris, for which he was awarded the $25,000 Orteig Prize in 1927.

The rules of the contest are simple. Participants must carry three people and must fly to an altitude of 328,000 feet (62.5 miles) above Earth before returning safely. They must then repeat the same flight with the same ship/plane/rocket within two weeks.

Since the X Prize was announced in 1996, more than twenty teams from seven countries have registered to compete. Aviation pioneer Burt Rutan, the first person to fly nonstop around the world in 1986, may be closest to winning the prize — thanks to a winged rocket (above), dubbed *Space Ship One*, that Rutan's company has developed. During a recent test flight over the Mojave Desert, the rocket soared to a height of 211,400 feet.

No matter who wins the coveted prize, the privatization of space will mark a profound change in our spacefaring activities. Whether it is for tourism or for the exploitation of natural resources in space, private capital is going to play a key role in spaceflight in the twenty-first century.

Because the distances are so vast, travel beyond Mars today is not a realistic prospect for human beings, at least not within the next hundred years. We might have a human rendezvous with an asteroid, but that would be as either a stunt or an accident. And there is some rationale for a return to the Moon. A base on the far side of the Moon that would manage a series of lunar-based telescopes is a real possibility. Such an array of telescopes would have its own unique set of problems, but in the long run the logistics involved in managing the equipment could prove less complicated than managing current low-Earth-orbit telescopes like *Hubble*. There has also been speculation about having resort hotels both in low Earth orbit and on the Moon. There is some prospect that one of these could be in place by century's end.

Sometime in the next century, we may want to send human explorers to one of the moons of Jupiter, but an extraordinarily compelling reason will be needed for that. Nothing less than the discovery of some life form larger than microbes would justify an expedition into the inhospitable domains of Jupiter.

The discovery of an Earth-like world around a nearby star would certainly prompt an interstellar spaceflight. First we would send a robot to perform reconnaissance, then perhaps human settlers would follow. (Of course, the hubris here is that if the extrasolar planet is habitable, it's most likely *already* inhabited!) For now, there is no way to reach even the nearest stars within a single human lifetime — or even thousands of generations of human lifetimes — given the state of current propulsion technologies. New means of propulsion must be developed.

> We must sail sometimes with the wind and sometimes against it,
> but we must sail, and not drift, nor lie at anchor.
>
> — Oliver Wendell Holmes

The Return of the Clipper | In 1872, the clipper ship *Thermopylae* left

Shanghai for London with a cargo of tea. Sailing in company with her was a second clipper, the *Cutty Sark*, also loaded with tea and bound for London. After racing each other for two weeks, *Cutty Sark* lost her rudder and was delayed for repairs. When she finally arrived in London, *Thermopylae* had already been there for seven days. Such races were common in the latter half of the nineteenth century, when clipper ships were both the ultimate expression of sailing technology and its last gasp.

It seems appropriate that a new age of sail should dawn from one of our planet's oceans. Sometime in 2004, *Cosmos 1*, the first solar sailing spacecraft, will begin its maiden voyage when it is launched into space aboard a converted intercontinental ballistic missile fired from a Russian submarine under the Barents Sea. This little sailing vessel, running without an engine, will rely solely on the pressure of photons striking its five-micron-thin mirrored Mylar sails to propel it forward.

Solar sailing is similar to maritime sailing in that the angle and size of the sail affect speed and performance. However, sails used on a boat often depend as much for power on the difference in air pressure between the front side of the sail and the back (like the lift of an airplane's wing) as they do on the push of the wind on the backside. A spacecraft using solar sailing can only "run before the wind," relying strictly upon the push of sunlight. For control, *Cosmos 1* can rotate its sails like the rotors on a helicopter. If it wants to sail away from the Sun, it positions its sails perpendicular to the direction in which it wants to go; if, on the other hand, the edges of the sails are pointed toward the Sun, a solar sailing vessel will fall back toward it, thanks to gravity. It should be noted here that *Cosmos 1* will not rely on the solar wind, which is a stream of particles, not photons. These particles will pass through the sails totally unencumbered.

If it works, *Cosmos 1* will have the distinct advantage of not having to carry any fuel. Acceleration is slow but, unlike chemical rockets, which can accelerate for only as long as they have fuel, a solar sail can accelerate continuously. If a solar sail were launched at the same time as NASA's planned *New Horizons* mission to Pluto, the solar sail would be left in the dust. But because the solar sail would continuously accelerate, it would eventually catch up, passing the *New Horizons*

probe roughly halfway to its destination. The *New Horizons* probe should take nine years to reach Pluto, while the solar sail would require only five. Another advantage of a solar sail is that it could make non-Keplerian orbits — that is, if the spacecraft wanted to hover in one spot, it could do so by balancing the pull of gravity with the push on its sail. This means the spacecraft could loiter in one spot and make long-term observations, or perhaps guard a specific patch of the solar system.

The disadvantage of a solar sail is that, beyond Jupiter, the sunlight grows appreciably dimmer. With fewer photons available to push it, the sail becomes less effective. One possible solution is to fill the sail with laser light. A laser could be based somewhere beyond Earth and be used to drive solar sailing ships efficiently — even to nearby stars. A second disadvantage is that a sail with a large surface area is needed to carry a small payload. Consequently, scientific instruments and spacecraft components must be miniaturized.

If successful, *Cosmos 1* would validate a whole new technology and could well bring back the great age of sail — this time, on the wider horizons of inter-planetary space.

Star Tech | The original *Star Trek* television series inspired a generation of kids.
(Myself included. I not only wanted to be like Mr. Spock, I wanted to *be* Mr. Spock. When I was in the third grade, I used to pinch the tops of my ears so that they would become pointed, just like his.) Perhaps the cult favorite also had some role in spurring the development of two new propulsion technologies — antimatter drives and ion drives.

In fact, an ion-drive spacecraft has already flown. On October 24, 1998, NASA launched a little motorboat of a spacecraft called *Deep Space 1*, the first

Today's exotic new propulsion technology may be tomorrow's common means of travel. Here, a solar sailing spacecraft similar to the Planetary Society's *Cosmos 1* is shown approaching Mars. Such spacecraft could prove useful for shipping supplies, cargo and scientific payloads to a base on the Red Planet.

Technicians at NASA's Jet Propulsion Laboratory prepare *Deep Space 1*'s xenon ion engine for testing in a vacuum chamber. The ring visible at the right side of the engine is its exhaust nozzle.

mission in its New Millennium program to test exotic new means of propulsion. The drive in *Deep Space 1* worked by ionizing xenon gas — that is, by putting an electrical charge on the xenon's atoms. The gas was then spun up electrically and fired out of an exhaust nozzle, pushing the spacecraft forward. Like solar sails, ion drives accelerate slowly. As a result, they lack the whoosh of a chemical rocket — but, like a solar sail, they create a steady acceleration that ultimately allows a craft to overtake a chemical rocket while using far less fuel. On *Deep Space 1*, the ion drive ran remarkably well.

Although its primary purpose was to be a test bed for ion-drive technology, *Deep Space 1* rendezvoused with Comet Borelly. Flying right through the comet's gaseous outer layers, the spacecraft sent back some amazing images of the icy comet's active nucleus — including a jetting geyser. *Deep Space 1* was finally decommissioned in December 2001, but its successful demonstration of ion technology proves that a better means of moving around space is already at hand.

An Explosive Possibility | In the 1950s the renowned physicist Freeman Dyson and his colleagues built a series of test models for a spacecraft that would use exploding atomic bombs as its means of propulsion. Calling their endeavor Project Orion, Dyson and company envisioned a craft that would boast a thick shield attached to a massive shock absorber built into its tail. To push forward, the craft would set off a series of atomic explosions. The shield would absorb the radiation and act as a pusher plate, with the shock absorber smoothing out the ride. The idea is roughly akin to a piston being pushed by exploding gasoline inside the combustion chamber. Dyson's test models were non-nuclear, and were designed to study the viability of the pusher-plate concept.

The idea of using atomic explosions as a means of propulsion dated back to the Manhattan Project, the top-secret program to develop the world's first atomic bomb. Two scientists, Stanislaw Ulam and Cornelius Everett, first proposed it, and the Defense Advanced Research Projects Agency — the same body later responsible for developing the Internet — subsequently took over the project.

With the signing of the Limited Test Ban Treaty in August 1963 — which banned nuclear-weapons testing in the atmosphere, in outer space and under water — almost no further development of Project Orion could occur. In fact, the exercise was illegal under the terms of the pact. Stymied by the treaty and eclipsed by the Moon Race, Project Orion lay dormant for the next thirty years.

In the mid-1990s, Penn State University physicist Gerard Smith, who had been thinking about the problem of overcoming the vast distances of interplanetary and interstellar space, proposed a spacecraft employing matter/antimatter annihilation as the primary means of propulsion. When an atom of matter comes into contact with one of its antimatter counterparts, the two are completely converted into energy. In other words, they're annihilated. This is the most efficient means there is of converting matter into energy, far more than chemical combustion in rockets or nuclear engines.

Smith's blueprint for a spacecraft named *ICAN-II* (an acronym for Ion Compressed Antimatter Nuclear propulsion) is reminiscent of Dyson's original Project Orion. Both ships require a pusher plate and a shock absorber, and both *ICAN* and Orion share the danger of a potential erosion of the pusher plate. Yet there are some problems unique to each design.

Unlike *ICAN*, Orion's use of atomic energy makes it "dirty." The thought of lofting atomic bombs into space — or the fissile material needed to create them — still makes a lot of people nervous. It's also still illegal. On the other hand, *ICAN*'s antimatter fuel is hard to make, even though very little of it would be needed. One hundred milligrams of antimatter could produce enough energy to equal the propulsive energy of the space shuttle. One kilogram (2.2 pounds) of antimatter could send a probe to the Oort Cloud.

The trouble is, the current production of antimatter at Chicago's Fermilab

In this dramatic illustration, Gerard Smith's *ICAN-II* spacecraft heads toward the Orion Nebula, fuelled by the pure energy created during matter/antimatter annihilation.

and Switzerland's CERN doesn't exceed ten nanograms each year. Given that one nanogram equals one billionth of a gram, it would take a considerable amount of time — millions of years — to accumulate a significant supply of antimatter. *ICAN* must also make use of magnetic "bottles" to store the antimatter: these bottles are, in fact, magnetic fields that keep the antimatter from touching any matter. The reliability of these bottles must never be less than one hundred percent, since any leakage would likely blow up the entire spacecraft. That's worth bearing in mind when you think of all the problems we face today in space — the leaky toilets, the uncoiling tethers, the stuck solar panels, the wayward antennae and the ever-present demon known as human error.

The obstacles are daunting but not insurmountable. Robert Frisbee heads up the Jet Propulsion Lab's Advanced Propulsion Office, a group that studies exotic means of spaceflight. They're searching for ways to move a spacecraft at speeds of up to one-tenth of the speed of light using antimatter fuel and laser light. "When my friends ask what I do, I sometimes reply tongue-in-cheek that I do government-funded science fiction," said Frisbee. "But then so were trips to the Moon fifty years ago."

Keep Looking Up | For the foreseeable future, we will have to conduct deep-space exploration with increasingly powerful, more exotic forms of space telescopes. Several telescopes now on the drawing board will be optimized to hunt for planets and to probe the Universe as it appeared during a time when galaxies were just forming.

The key word in new space-telescope design is interferometry — a technique that allows the light from several telescopes to be combined into a single image. This means that two or more small telescopes can gain the power of a much larger one by combining their incoming light. And their total power would be more than the sum of their parts. For example, a pair of telescopes 21 meters in diameter set 500 meters apart will not behave like a 42-meter telescope, but like one with a diameter equal to the distance from each other — 500 meters across.

This technique has been demonstrated time and again in the radio domain; consider the famous scene in the movie *Contact* that shows Jodie Foster's character sitting in front of a row of radio dish antennae. These antennae are working in unison as if they were a single unit, thanks to the radio version of interferometry. Only now are we able to apply this technique to the optical portion of the spectrum, where much greater precision is required to make it work.

Three upcoming missions will push the limits of interferometry: the *Space Interferometry Mission* (*SIM*), the *Terrestrial Planet Finder* (*TPF*) and *Constellation-X* (*Con X*). *SIM* and *TPF* will have their optical mirrors arrayed on a bench, while *Con X* — an X-ray telescope designed to succeed the *Chandra X-ray Observatory* — consists of four small spacecraft flying in formation. *LISA* is another space observatory that will rely on formation flying, although it is not an interferometer.

The *Hubble Space Telescope*'s immediate successor has been christened the *James Webb Space Telescope* (*JWST*), in honor of an Apollo-era NASA administrator. This telescope will employ a large, segmented mirror — two and a half times the size of *Hubble*'s — to concentrate on infrared astronomy. (*Hubble* was optimized for optical and ultraviolet astronomy.) To some extent *JWST* will also complement the *Spitzer Space Telescope Facility*, an orbiting mission scheduled to end by 2009. The *James Webb Space Telescope* is specifically designed to probe the formation of the first stars and galaxies and to measure the geometry of the Universe and the distribution of dark matter. Unlike *Hubble*, *JWST* will not be serviceable in space, since it will orbit Earth beyond the reach of the shuttle — in what is called a Lagrange point, roughly one million miles from Earth. (Lagrange points are positions around Earth and the Moon where the gravitational pull is equal between the two, so a spacecraft can be "parked" there).

Orbiting telescopes like these, as well as superlarge, superpowerful ground-based ones, will contribute to our overall understanding of the Universe. And although actual travel to the stars may not occur anytime soon, our imaginary flights through the cosmos will carry on — thanks to the findings of these powerful new telescopes.

CHAPTER SEVEN # The Big Picture

If we do discover a complete theory [of the Universe]...
it would be the ultimate triumph of human reason — for
then we would know the mind of God.

— Stephen Hawking

As we travel deeper into space, we also travel farther back in time. Because everything in the cosmos is so far away, we see it as it *was*, not as it *is*. The Moon is 1.3 light-seconds away, which means you see the Moon as it was 1.3 seconds ago, not as it is at this instant. The Sun is roughly 8.33 light-*minutes* away. The famous Pleiades star cluster — easily spotted at night by the naked eye because of its distinctive dipper shape — is 380 light-*years* from Earth.

As noted in chapter four, the light we see today left the Pleiades about the same time as the pilgrims landed on Plymouth Rock. Similarly, the light that reaches us from the Orion Nebula, a star-forming region 1,600 light-years away, began its journey to Earth as Rome was being sacked by the Visigoths. We see the Magellanic Clouds as they were 169,000 years ago, when bands of Ice Age hunters stalked herds of woolly mammoth on the Eurasian Steppe. If Gerard Smith's *ICAN-II* spacecraft ever found its way to the Magellanic Clouds, any problems it had to report to Mission Control in Houston would take 169,000 years to reach Earth — and that long again for their reply.

To put it another way, the light we see today is a kind of fossil record of the history of the cosmos.

We are now one hundred million light-years from Earth. We can see that our Local Group is falling toward the Virgo Cluster, a gathering of about 150 large galaxies and more than a thousand dwarf satellite galaxies some sixty million light-years away. At the heart of Virgo is the massive elliptical galaxy M87, home to a

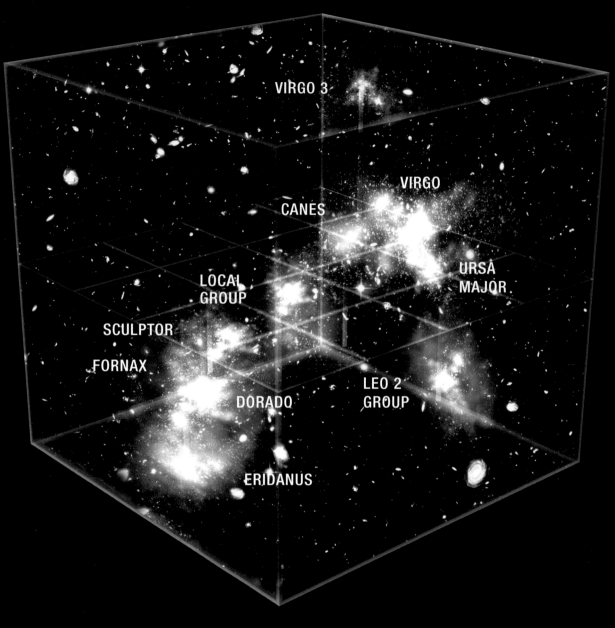

VIRGO 3

VIRGO

CANES

LOCAL GROUP

URSA MAJOR

SCULPTOR

FORNAX

DORADO

LEO 2 GROUP

ERIDANUS

This is a view of the Virgo Supercluster, which contains our Local Group (the Milky Way and Andromeda) and the Virgo Cluster — the closest chunk of large-scale real estate. A supercluster is a group of galaxy clusters, all bound by dark matter. At this large scale, the effects of dark energy may also be detected.

No one really knows what the Great Attractor is, but we can see its effect by the massive infall of thousands upon thousands of galaxies and galaxy clusters. One possible explanation is that the Great Attractor is a gargantuan cloud of dark matter herding these galaxies with its gravitational pull. That cloud is shown in pink and lavender in the image above.

three-billion-solar-mass black hole. Our Local Group is rushing toward Virgo at about 370 miles per second. Unless dark energy prevents this, the Local Group will one day be cannibalized by the Virgo Cluster. Beyond Virgo is the Coma Cluster, 80 megaparsecs from Earth (that's more than 280 million light-years away). The Coma Cluster, Virgo and our Local Group are all roughly aligned in what astronomers call the supergalactic plane. The galaxies in this plane form the Local Supercluster, with the Virgo Cluster at the center.

The convergence of our Local Group with the Virgo Cluster is but a small eddy in the heavens compared to the motions of the Local Supercluster. Virgo, Coma, another cluster called Hydra, the Centaurus Supercluster and several other galaxy clusters appear to be flowing toward a dark, unseen mass known to astronomers as the Great Attractor. The pull is so great and extends so far that it can be thought of as a mighty river in space.

The nature of the Great Attractor, which was first noticed in the 1980s, is not fully understood. Astronomers believe that a huge concentration of mysterious dark matter there is responsible for the massive infall of galaxies. Dark matter itself is an utter mystery. It wasn't until the late 1970s that we became aware that what we were seeing with our telescopes wasn't all that was there.

In 1976, Vera Rubin, an astronomer at the Carnegie Institute, was analyzing the way galaxies spin. She expected that their rotational velocities would be consistent with Newton's laws. Surprisingly, she found that the outer arms of spiral galaxies spin faster than they should. Rubin concluded that their faster spin must be due to the gravitational influence of something unseen. Rubin's analysis drew attention for the first time to the presence of a new, strange, dark matter.

Conjecture about what dark matter is has ranged from the now-disfavored theory that it is an ocean of tiny snowballs and ice pellets to the exotic one about Weakly Interacting Massive Particles (WIMPs). No consensus exists, but the WIMP idea has a lot of appeal among researchers — in part because we already know about neutrinos, elementary particles that can pass through Earth totally unhindered and are extraordinarily difficult to detect. It's a good bet that physicists have yet to

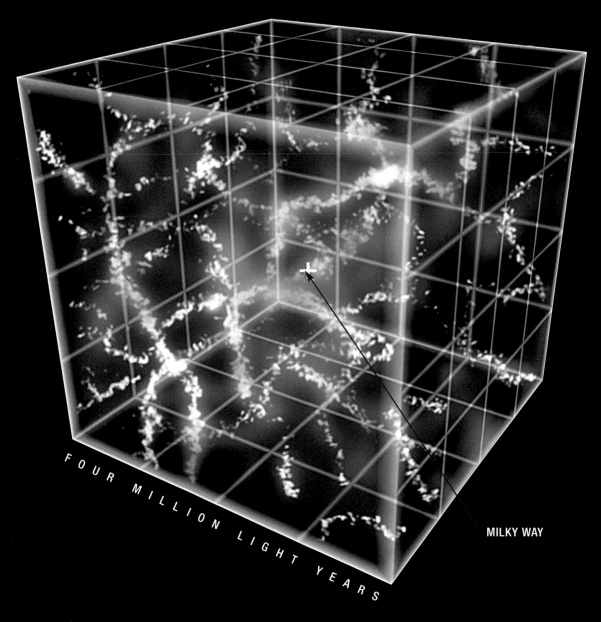

FOUR MILLION LIGHT YEARS

MILKY WAY

Derived from the work of Geller, Huchra and de Lapprent, as well as T. Theun at the Max Planck Institute, this chart shows the web-like topography of large-scale structure in the Universe. We begin to see here the mature forms of structures that began as clumps in the *WMAP* image back on page 17. Gravity, dark matter and dark energy all play a role in sculpting the hollows and clumps of galaxies we find on such scales.

describe the full gamut of subatomic particles, and that as-yet-undetectable WIMPs are lurking out there.

As we continue our outbound journey, we encounter a strange phenomenon — essentially, a mirage in space called the Einstein Cross. A central galaxy is surrounded by the squeezed images of other galaxies much farther away, in direct line of sight with this first galaxy. The Einstein Cross is an example of galactic lensing, an effect that proves that gravity bends space. Astronomers use such rare occurrences of lensing to see farther back in space-time than even the *Hubble Space Telescope* could otherwise see. *Hubble* first spotted the Cross in 1990.

We are now billions of light-years from Earth. We see the Virgo Supercluster receding in the distance and can grasp the large-scale structure of the entire Universe. Before the 1980s, it was assumed that galaxies were spread roughly uniformly across the Universe. In 1983, Robert Kirshner, Gus Oemler, Steve Shectman and Paul Schechter found evidence of a Great Void in the constellation Bootes. Measuring some 100 megaparsecs across (that's 326 million light-years), the Great Void was the first hint that galaxies were not evenly distributed.

Then, in 1985, at the Harvard-Smithsonian Center for Astrophysics, Margaret Geller and John Huchra, along with Valerie de Lapprent, released their first surveys of large-scale galactic structure. These were great, three-dimensional

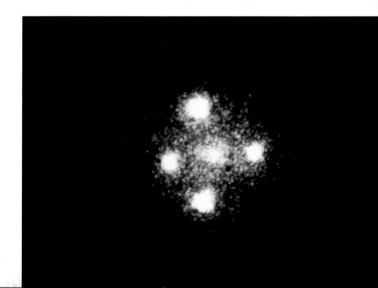

This spectacular image of the Einstein Cross, captured by *Hubble*, proves the physicist's theory that massive objects curve space around them. Two galaxies are in the same line of sight — the one in the middle, and the four that flank it. The four flanking galaxies are actually a single galaxy. Light from the more distant galaxy is split and bent around the central one, creating a mirage in space of four surrounding galaxies.

Mirages in Space

There's something a little spooky about gravitational lenses. The images they send back to Earth are akin to mirages in space. And as with any mirage, objects are not where they appear.

Gravitational lenses were first described by German astronomer Johann George von Soldner at the Berlin Observatory in 1804, but the phenomenon was never investigated until Einstein revived interest in them in 1911. His prediction of gravitational lenses came from the idea that massive objects curve the space around them. Light rays from a distant glowing object (A), traveling through curved space, would be forced to bend (B) and magnify. This refraction of the light is just like that of a glass lens. If the light source and the lens are perfectly aligned with the observer (C), and if the lensing object is perfectly symmetrical, the lensing effect would be a ring. But since the alignments are rarely perfect and the lensing objects never symmetrical in their mass distribution, the light is broken into four images (D) arrayed in a cross pattern.

In 1937, American astronomer Fritz Zwicky suggested that such gravitational lenses could provide a means of studying extremely remote objects, but the first gravitational lens was not observed in nature until 1979. The famous Einstein Cross (page 153) was the first to be imaged by the *Hubble Space Telescope*. After its servicing mission in 2000, *Hubble* also captured the dramatic image of the galaxy cluster Abell 2218 (left) — some two billion light-years from Earth. The cluster is so massive and so compact that its gravity bends and focuses the light from galaxies that lie behind it. As a result, multiple images of these background galaxies are distorted into long faint arcs — a simple lensing effect analogous to viewing distant street lamps through a wine glass.

At a recent conference held at the Stanford Linear Accelerator Center near San Jose, California, astrophysicist Alexandre Refregier from the Commissariat à l'Energie Atomique in France described how such lenses can be used to chart dark matter. By making statistical measurements of the strength of the lens, we might be able to map the distribution of dark matter at various scales, make the first three-dimensional topographical maps of individual dark-matter clouds and measure the evolution of large-scale structures from extremely early epochs. It may even be possible to locate gravitational lenses in places where the lensing object is not at all obvious or even visible. Such lenses would most likely be clouds of dark matter unaffiliated with any particular galaxy cluster.

At the dawn of the twenty-first century, we are finding that gravitational lenses are far more than just a mysterious special effect. They are becoming tools for studying dark matter, and for seeing where no telescope could see before.

D

D

B

A

B

C

D

D

pie slices of the sky that initially extended out to about two billion light-years. Some 1,500 galaxies were plotted in the three-dimensional coordinate space relative to the Milky Way. For the first time, we could see that galaxies were distributed along thin filaments around massive voids. The Universe's structure was likened to a sponge, with thin fibers of material surrounding empty spaces. To scientists, this strongly suggested that vast areas of dark matter governed the distribution of galaxy clusters.

Since then, Geller and Huchra have released their Century Survey, which carries the plotted locations of galaxy clusters out to a distance of three billion light-years. Such surveys may one day reach out some thirteen billion light-years from Earth, taking us back to the Dark Ages of the Universe's formation.

We are now five billion light-years from Earth. Light began leaving objects this far away at a time when our Sun and Earth were just forming. On the molten ball we would one day call Earth, there were no plants, no animals, no atmosphere, no oceans — only intense radiation from our newborn Sun, and constant bombardment from meteors and asteroids.

It bears repeating here that our journey outward runs our clock backward. Around us, the core regions of individual galaxies begin to flare with the light of quasars and galaxies with active nuclei. We now pass through the quasar epoch and enter the Dark Ages, a time about which very little is known. We are now seeing how galaxies and quasars first began forming inside giant primordial clouds.

BIG BANG

The Arrow of Time. This idea, championed by Eric Chaisson, provides a framework for organizing our understanding of the evolution of the Universe. Here we see the entire history of the Universe — from the Big Bang 13.7 billion years ago, at left, to the present moment. The evolution of space-time is shown in the horn shape (A). The curve in the horn represents the acceleration in the rate of our Universe's expansion, with each blue ring marking off one billion years. Inside the horn, we can see the evolution of large-scale structure. The purple gradient in the bar behind the horn (B) charts the demise of energy and the corresponding rise in dark energy. The history of the Universe can also be marked by the change from simple forms of matter to complex ones (C). The red arrow at the bottom indicates the direction in which time flows — always from the past, through the present and into the future.

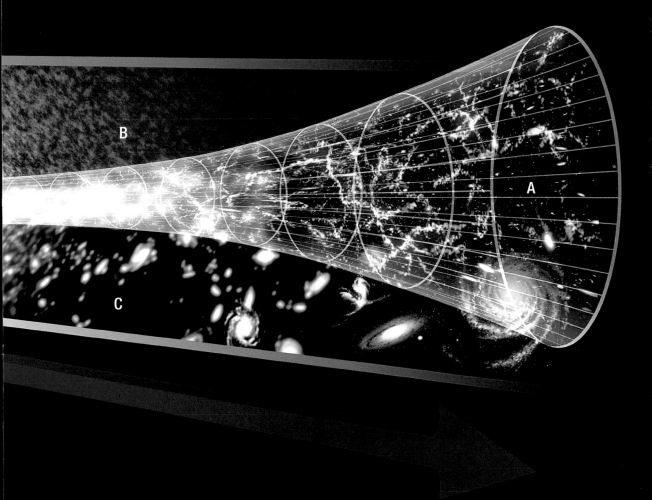

The first stars began to form approximately two hundred million years after the Big Bang. These were no ordinary stars. Large, energetic and short-lived, they burst to life and then exploded with such rapidity that they have been described as the ultimate fireworks show.

As these early stars were born, great shards of gas, dust and stars began to congregate into ever-larger agglomerations. In the center of these clouds, stars began to collide and form black holes; these black holes became ever greater in mass and strength, sucking in surrounding stars with machine-gun-like rapidity, ultimately taking on supermassive proportions. This scenario may well describe the birthing not only of the first black holes but also of the first galaxies.

Ahead lies decoupling. This is the moment when the entire Universe underwent a phase transition. We see such transitions every day when, for example, water changes into ice or steam. For the Universe, this transition was from an opaque environment in which energy prevailed, to a transparent environment in which matter prevails.

The *WMAP* image on page 17 represents a back wall beyond which we are presently unable to probe. In the photo collage above, we see the galaxies of the *Hubble* Ultra Deep Field congealing from the blobs that characterize the *WMAP* image. In reality, those blobs represent gigantic clusters of galaxies — not individual ones.

We are now approaching the frontier of time and space. We are some thirteen billion light-years from Earth, in a time when our Sun and its planets didn't exist. As if we were watching the ultimate time-lapse movie, we see the entire Universe as a roiling, seething realm of immense activity. Since we are traveling backward in time, we see everything around us moving backward. Instead of expanding, we see the Universe collapsing on itself. The Universe changes from cool and transparent to an opaque, white-hot realm — a sea of loose atoms that in turn dissolve into seething, unbound particles.

This artwork provides a "God's Eye" view of the moment of creation, also known as the Big Bang.

We have now traveled so far back in time that we are at the primal moment of creation, the Big Bang. Time itself is at an end point and we can travel no farther back. Human history is long gone, thirteen billion years behind us — or, you could say, ahead of us. We now see a single white dot, glowing, the Primordial Kernel, and we can only guess at what caused it to expand in the first place. Is the Universe some quantum fluctuation or the work of God? For the moment, the answer to this question lies beyond the reach of science, but there are some who have sought to introduce the idea of design into the cosmological debate.

The Anthropic Principle | Perhaps more philosophy than science, this principle states that the Universe is fine-tuned in every way to permit the rise of life, and that the rise of higher forms of biology must therefore be its purpose. Change any single parameter of the Universe — the amount of energy, the number of dimensions, the amount of matter, the rate of expansion, the age or any given

physical law — and the Universe ceases to be suitable for life. Our existence demands that biological constraints be applied to the "miraculous" balance of all of the physical parameters of the Universe.

Many scientists are uneasy about the Anthropic Principle because of its theological implications. Advocates contend that it is handy because, if nothing else, it sets limits on the range of our investigations. Metaphorically speaking, it becomes a kind of speed limit that disallows theories of a Universe inhospitable to living things.

Critics say the Anthropic Principle is an anti-Copernican sleight of hand that places humans back in the center of the Universe. Further, the concept presupposes its answer, so there is no way to disprove it. In any case its calculations are, at least for the moment, imprecise.

This is a telling point. One of the key variables that the theory relies on is the so-called cosmological constant — an extra term in Einstein's equations for general relativity that describes the density and pressure associated with "empty" space, a kind of vacuum energy. This was the force that kept the Universe expanding, especially during the crucial period when galaxies were first forming. Einstein called the cosmological constant his greatest mistake, but we now see that he was correct. The discovery of dark energy validated Einstein's original idea and, while its precise value still escapes us, from *WMAP* we know that it accounts for seventy-three percent of our Universe. Debate over the Anthropic Principle is closely intertwined with the resolution of this cosmological constant.

The Brane Game | One question that may not be totally beyond our reach is

whether our Universe is the only one. This question at first glance is a semantic red herring, for the definition of a Universe is all that is, ever was and ever will be.

In 2001, cosmologists at the Space Telescope Science Institute put forward an alternative idea for the origins of the Big Bang. Suppose that the four dimensions we experience every day — the three spatial dimensions of height, width and depth, along with the fourth dimension, time — are not all there is. What

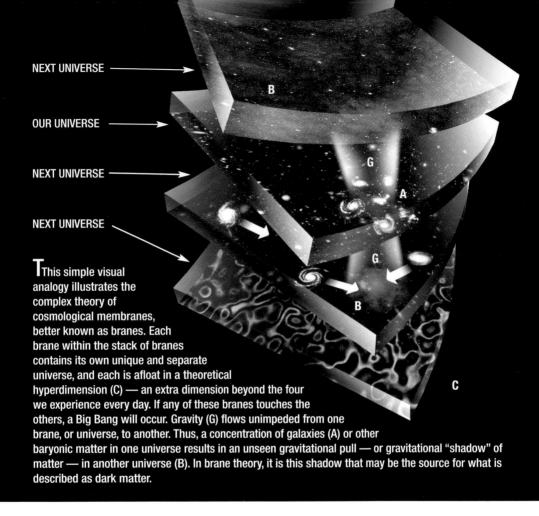

NEXT UNIVERSE

OUR UNIVERSE

NEXT UNIVERSE

NEXT UNIVERSE

This simple visual analogy illustrates the complex theory of cosmological membranes, better known as branes. Each brane within the stack of branes contains its own unique and separate universe, and each is afloat in a theoretical hyperdimension (C) — an extra dimension beyond the four we experience every day. If any of these branes touches the others, a Big Bang will occur. Gravity (G) flows unimpeded from one brane, or universe, to another. Thus, a concentration of galaxies (A) or other baryonic matter in one universe results in an unseen gravitational pull — or gravitational "shadow" of matter — in another universe (B). In brane theory, it is this shadow that may be the source for what is described as dark matter.

if our four-dimensional Universe floats in a fifth dimension — an overarching metadimension that has our familiar four dimensions bound together on what scientists call a cosmological membrane, or brane for short? This brane theory suggests that the Universe we can see, know and measure exists on one of these branes. Or, put another way, time and space are embedded within the fabric of a brane. There may be other parallel branes floating in this metadimension, and if any two branes collide with each another, a Big Bang will occur.

One consequence of a multi-brane Universe is that gravity from baryonic matter (the stuff we experience every day) embedded on one brane can influence the matter on another — one casts a kind of gravity "shadow" on another. This is because gravitons — the virtual quantum particles that convey the force of gravity — can flow unimpeded between branes. Concentrations of matter in one brane will cause matter in another to pool under its shadow. The effects of that shadow may be what we interpret as the presence of dark matter. Could it be that the Great Attractor is one of these shadows from a neighboring brane?

The beauty of the brane theory is that it allows an infinite number of universes to exist simultaneously on different, parallel branes, none of which is beholden to the Anthropic Principle — except, of course, for one. Ours. If other universes live by different rules of physics, then there must be something that causes variation among these — a DNA for a cosmos. Anything goes. Almost.

At the End | Although we may not know why our Universe exists, we are beginning to understand how it came about. But what will be our cosmic fate?

Hubble's greatest discovery to date is the acceleration of the Universe. This means that the Universe is not only expanding but also doing so at an ever-increasing rate of speed. One day, this cosmological expansion will be so great, the distances so large, that no light from any other galaxy could ever reach us. One by one we will see galaxies fade into the night. Astronomers in some future time might believe their galaxy is the only one. They will debate the extent of the Universe, just as Curtis and Shapley did in the 1920s, but, with no observational evidence of the myriad galaxies beyond the cosmic horizon, their version of that debate will reach different conclusions. Our galaxy, they will believe, is all there is.

Will our Universe expand forever? If these latest findings from *WMAP* and *Hubble* hold true, then the answer is yes. This expansion will eventually become so great and so fast that, billions of years from now, the Universe will begin to freeze out. All the loose gas within our galaxy will have been swept into stars, and the stars will finally all burn out. The Milky Way will darken as the ambient temperature of the Universe — itself cooling off since the Big Bang — drops and finally freezes out.

Some theorists have speculated, however, that the last moments of the Universe will not end as a sleepy fade to black. Instead of freezing out, the expansive force of dark energy will grow so strong that galaxies will not only be pushed away from each other but be torn apart as well. Then stars and planets will fly apart and in the last milliseconds of the Universe, even atoms will be torn asunder. This scenario has been dubbed the Big Rip.

Whether the end comes by freezing out or by the explosive power of the Big Rip, somewhere along the way, the Universe will no longer provide hospitable conditions for life. Thus the awakening of consciousness in our cosmos is temporary and ultimately doomed.

Will this be the result of a new, previously unknown force of nature, a kind of antigravity that has come to be known as dark energy or, as Lawrence Krauss calls it, quintessence? And what role does dark matter play here? Only time and further work will tell.

Much of what we know about the fate of the Universe has come from the *Hubble Space Telescope*. Less certain, however, is the fate of this extraordinary instrument. Eventually, *Hubble* will be decommissioned and sent on a fiery plunge into the Pacific Ocean. The original plan to retrieve it for the halls of the Smithsonian was dropped because of safety concerns in the wake of the *Columbia* disaster. These concerns were again cited by NASA when the space agency decided not to proceed with *Hubble*'s last scheduled shuttle servicing mission. Instead, NASA has proposed that a robot be sent to rendezvous with *Hubble* and perform the minimum tasks needed to keep the orbiting telescope aloft. As this book goes to press, the various designs proposed for a servicing robot are being evaluated. If this idea is successful, robots may one day service the *Chandra X-ray Observatory, Spitzer Space Telescope* and other vital missions not designed originally for refurbishment. And so, along with the upcoming *JWST, Kepler, LISA, Constellation X* and a host of future missions, the work will continue.

Epilogue

We shall not cease from exploration
And the end of all our exploring
Will be to arrive where we started
And know the place for the first time.

— T.S. Eliot,
from "Little Gidding," *The Four Quartets*

But why bother? | After all, there are so many other things that require our immediate attention, all the time: What's for dinner? Where's the rent? Paying the bills, fixing the car, going to work, going to school and on and on. If that's not enough, we have wars and famine and pollution and politics run amok. Crime, terror and corruption know no bounds. And then there's *always* the weather.

With so much swirling around us in our daily lives, why bother with things like the fate of the Universe or where it all came from? How is exploring the cosmos or sending people to Mars relevant to our daily lives? As the astronomer Steve Maran put it at a recent conference hosted by the National Academy of Sciences in Washington, "How many of us know someone who died from a supernova?" The answer, of course, is no one.

Astronomy is not going to solve crime or make a better gadget. In fact, it has been sixty-five million years since the last time the cosmos had any significant impact upon the daily lives of the denizens of Earth. And that, by the way, is a good thing.

Could it be that we are simply trying to locate ourselves, perhaps — as Margaret Geller put it — fill out our address in long form? To add, after our street, city and country, the bigger details, like Earth, Our Solar System, The Local Bubble, Orion Arm, Milky Way Galaxy, The Local Group, Virgo Supercluster, Our Universe's Brane? Where would it end?

Astronomy was once the tool of priests and kings and farmers, but it has very little practical value today. And that's all right. Man has always spent time on things that are of no practical value. How is a Beethoven symphony going to solve crime? When is the Mona Lisa going to fix my car?

The point is that astronomy is more akin to art or music. It tells us something about ourselves, something deep and primal. Like all science, astronomy seeks to find the limits of knowledge — for those are the limits of the human mind. Do the laws of nature converge, as Richard Feynman put it, or do nature's mysteries only widen, as Einstein supposed? Science, like art, changes the way we see the world.

On a warm summer night, a father and his son peered through a little telescope at a fat, full Moon. As they took turns at the viewfinder, the cares of the day were lost in the wonder of the cratered lunar surface. The crickets buzzed and the frogs croaked as the smell of magnolia wafted through the air. Wave your hands, son, and skip in the moonlight. The mysteries of the cosmos are yours to solve.

Captured on a "canvas" in interstellar space, this February 2004 image from the *Hubble Space Telescope* bears remarkable similarities to Vincent van Gogh's famous *Starry Night*, with its bold whorls of light sweeping across a raging night sky. *Hubble*'s "starry night" is the latest view of an expanding halo of light around a distant star located about 20,000 light-years away from Earth, at the outer edge of our Milky Way galaxy. The bold whorls here are never-before-seen spirals of dust swirling across trillions of miles of space.

(Overleaf) A remarkable double star cluster captured by *Hubble*.

Index |

References to images are indicated by italicized page numbers.

A

Abell 2218 (galaxy cluster), *154*, 155
Absorption lines, *120*, 121
Accretion disks, 96, 97, *97*, 98, *98*, 99
AGNs (Astral Galactic Nuclei), 104
Albedo maps, 52
Aldebaran (star), 55
Aldrin, Buzz, 31, *126*, 127
Alpha Centauri, 24, 74, 75, *75*, 87
Alvarez, Luis and Walter, 30, 31
Ames Research Center, 34
Andromeda, 116, 117, *117*, 118, 119, *119*, *122*, 123
Anglo-Australian Observatory (Australia), 85
Anthropic Principle, 159–60, 162
Antimatter drive technology, 141
Apollo 11, 127
Apollo 12, 66, *66*
Apollo 17, 21, 127
Armstrong, Neil, 72
Asteroid Belt, 40, *40*
Asteroids, 24, 30–31, 40–41, *40*, 156
Astrobiology, 58
Astronomical unit (AU), 52, 110
Atomic explosions, propulsion, 143, 144
Atomic fusion, 23, 27

B

Baryonic matter, 162
Beagle 2, 37
Bean, Alan, 66, *66*
Bell, Jocelyn, 95
Bell Labs, 17, 18, *18*
Bennett, Chuck, 16
Berlin Observatory, *154*
Beta Pictoris, 24, 90, *91*
Big Bang, 16, 18, 19, 100, 105, 119, 156, *156*, 157, 158, *158*, 160, 162, 163
Big Dipper, 125
Big Rip, 163
Black holes, 18, 95–97, *97*, 98, *98*, 99–100, 101, 102, *102–3*, 104–106, *105*, *106*, 107, 113–114, 158
Blueshift, 118–19
Bond, Howard, 109–10, 116
Bootes (constellation), 153

Brane theory, 160, *161*, 162
Brown dwarf star, 79, 80
Brown, Michael, 53
Buie, Marc, 52
Bunsen, Robert W., 120–21

C

Callisto (Jovian moon), 68
Cape Canaveral, 128, 131
Cassini Gap, 48
Cassini, Giovanni, 48
Cassini-Huygens probe, 48, *48*, 49
Cassiopeia A, 92, *92*, *93*
Cauchy Horizon (black hole), 102, *102–3*
Centaurus Supercluster, 151
Century Survey, 156
Cepheid Variable (stars), 118
Ceres (asteroid), 40
CERN (Geneva), 106–7, 146
Chaisson, Eric J., 6–9, 157
Challenger (space shuttle), 128, 131
Chandra X-ray Observatory, 15, 89, 90, 92, 99, 104, 105, 147
Chariots of the Gods (Erich von Däniken), 73
Charon (Plutonian moon), 50, *51*, 53
Chemosynthesis, 63
China, space program, 132–33
Civilian spaceflight, *137*
Clementine (probe), 31
Close Encounters of the Third Kind, 70
Color spectrum, 120, *120*
Columbia (space shuttle), 16, 128, 131, 163
Coma Cluster, 151
Comet Borelly, 143
Comet Hale-Bopp, 42
Comet Halley, 41–42
Comet Shoemaker-Levy 9, 43
Comet Wild 2, 42, *42*
Comets, 24, 31, 41–43, *42*, 124
Conrad, Pete, 66
Constellation-X, 147, 163
Contact, 137, 147
Copernicus, Nicolaus, 19, 22, 43
Cosmic Background Explorer (*COBE*), 17
Cosmological constant, 160
Cosmological expansion, 16, 19, 158, 160, 162–63
Cosmological membranes, 160, *161*, 162
Cosmos I, 139, 141
Curtis, Heber, 118, 162
Curved space, 153, 154, 156
Cygnus (constellation), 101
Cygnus X-1 (black hole), 101

D

Dactyl (moonlet), 41
Dark energy, 16, 107, 116, 149, *149*, 152, *152*, 157, *157*, 160, 163
Dark matter, 16, 114, 116, 147, 150, *150*, 151, 152, *152*, 153, 155, 162
Darwin, Charles, 19
Davis, Donald R., 32
Day the Earth Stood Still, The, 70
Decoupling, 17, 19, 113, 158
Deep Space 1, 141, 142, *142*, 143
Defense Advanced Research Projects Agency, 143
Deimos (Martian moon), 40
de Lapprent, Valerie, 153
Devon Island, 134
Daimandes, Peter, 137
Dicke, Robert, 18
Discovery (space shuttle), 130, *130*
Doppler effect, 78, 119
Drake Equation, 73
Drake, Frank, 73
Dust disk, 97, *97*
Dwarf galaxy, 116, 117, *117*, 125
Dyson, Freeman, 136, 143

E

Earth, 1, 15, *15*, 21, 22, *22*, 26, 27, 28–29, 30, 31, 32, *32*, 58, 59, 66, *76*, 77, 148
EGGs (Evaporating Gaseous Globules), 84
82 Eridani (star), 74
Einstein, Albert, 100, 101, 119, 154, 160, 165
Einstein Cross, 153, *153*, 155, 156
Electromagnetic spectrum, 18
Emission lines, *120*, 121
Endeavour (space shuttle), 128, *129*, 131
Epsilon Indi (star), 74
Eskimo Nebula, 92, *92*
Eta Carinae (star), 87, *88*, 89–90, *89*
Europa (Jovian moon), 45–46, *46*, 47, *47*, 64, 66–69
European Space Agency, 41, 98, 104
Evans, Ronald, 127
Event horizon (black hole), 97, 100, 102, *102*, *103*, 104, 105, 107
Everett, Cornelius, 143
Exoplanets, 79
Extinctions, mass, 30, 31, 86
Extrasolar gas giants, 77, 79
Extraterrestrial communication, 70–71

F

51 Pegasus (star), 77
Ford, Holland, 97
Fortuna Tessara (Venus), 29
47 Tucanae (star cluster), 81, 116
Fossils, 31, 35, 62, 65–66, 135
433 Eros (asteroid), 41
Frame-dragging, 100, 102, *103*
Fraunhofer lines, 120, *120*
Frisbee, Robert, 146

G

Gagarin, Yuri, 72
Galactic bulge, *112*, 113
Galactic collisions, 32, *32–33*, 40, 113,
    119, 124, *124*
Galactic lensing, 156
Galactic merger, 119, *122*, 123–25
Galaxies, 113–14, 118, 125, 151, 156,
    157, 158
Galilean moons, 43, *44*, 45
Galileo, 15, 43, 44, 48
Galileo (probe), 40, 45, 46–47, 49,
    58–60, *59*, 66
Ganymede (Jovian moon), 45, 68
Geller, Margaret, 153, 156, 164
Gas giants, 43
Gaspra (asteroid), 40
Gemini (constellation), 92
Genesis, Book of, 12
Geysers, 96, 98, *98*, 99, 143
Giotto (probe), 41
GI star cluster, 106, 119
Global warming, 30
Globular star clusters, 80–81, 114,
    115, 116
Goldilocks Zone, 75
Gould, Stephen Jay, 31
Gravastars, 107
Gravitation, 40, 45, 47, 67, 98, *98*, 99,
    100, 107, 113, 114, 139, 141, 147, 151
Gravitational lenses, 154–55, *154–55*
Gravitons, 162
Gravity spectrum, 18
Gravity waves, 18–19
Great Annihilator (black hole), 95–96
Great Attractor, 150, *150*, 151, 162
Great Dark Spot (Neptune), 50
Great Red Spot (Jupiter), 43, 50
Great Void, 153
Greenhouse effect (Venus), 28
Greenhouse gases, 30
Gusev Crater (Mars), 39, *39*

H

H II regions (nebulae), 110
Halley probes, 41–42
Hallucigenia, 62, *62*
Hartmann, William K., 31, 32
Harvard-Smithsonian Center for
    Astrophysics, 153
Hawking, Stephen, 100, 104, 107
Hawking radiation, 105–6, 107
Hayden Planetarium (NYC), 50, 52
HD 44594 (star), 24
HD 202206 (star), 79
Heliopause, 55
Hellas Basin (Mars), 37
Herschel, William, 92
Hester, Jeff, 81, 84, 86
Heterotrophic microbes, 64
Hewish, Antony, 95
Huchra, John, 153, 156
Hubble Deep Field, 125
Hubble, Edwin, 18, 118–19, 121, 125
Hubble Space Telescope, 15, 23, 24, 27,
    37, 43, 45, 48, 52, 80, 81, 82, 84, 85,
    86, 87, 89, 90, 91, 96, 98, 106, 111,
    118, 119, 124, 125, 128, 130, *130*,
    138, 147, 153, 155, 156, 162, 163
Hubble's Law, 121
Huygens, Christiaan, 48
Hyades (star cluster), 90
Hydra (star cluster), 151
Hydrobot, 68–69, *68*
Hydrothermal vents, 67–68, *68*
Hypernovae, 18
Hyperplanet, 80
Hyperthermophiles, 62

I

ICAN-II (spacecraft), 144, *145*, 148
Ida (asteroid), 40–41, *40*
Impact craters, 29, 45
Impactor, 32, *32*
Infrared astronomy, 147
Interferometry, 78–79, 146–47
International Space Station, 133–34,
    137
Io (Jovian moon), *44*, 45, 46, 59, *59*, 67
Ion-drive technology, 141, 142, *142*,
    143

J

Jet Propulsion Laboratory (NASA), 30,
    42, 46, 78, 142, *142*, 146
JIMO (*Jupiter Icy Moons Orbiter*), 68
Jupiter, 15, 22, *22*, 40, 43, *44*, 45, 46,
    46, 47, 49, 50, 59, 67, 138, 146–47
Juventae Chasma (Mars), *134–35*, 135
JWST (*James Webb Space Telescope*),
    147, 163

K

Kepler, Johannes, 22, 78
Kepler (telescope), 78, 79, 162
Key Hole Nebula, 23, *23*, *111*
Kirchhoff, Gustav, 120–21
Kirshner, Robert, 153
Kitt Peak National Observatory, 82
Krauss, Lawrence, 163
KT boundary, 30–31
Kuiper Belt, 47, 52, 54, *54*
Kuiper Belt Objects (KBOs), 50, 52, 53,
    53, 55
Kuiper, Gerard, 52

L

Lagrange point, 147
Laplace, Pierre-Simon, 101
Large Hadron Collider, 106, *106*
Large Magellanic Cloud, 86
Lemaître, Abbé Georges, 119
LGM–1 (pulsar), 95
Liller 1 (star cluster), 95
Limited Test Ban Treaty (1963), 144
LISA (*Laser Interferometer Space
    Antenna*), 18–19, 114, 147, 163
Local Bubble, 90
Local Group, 116, 118–19, 123–25,
    148, 149, *149*, 151
Local Supercluster, 151
Long-period comets, 55
Los Alamos National Laboratory, 107
Lowell, Percival, 34
Lunar Prospector (probe), 31

M

M15 (star cluster), 106
M16 (Eagle Nebula), 81, 82, *82–83*, 84, 87
M80 (globular star cluster), 114, *115*
M81 (spiral galaxy), 119, *119*
M87 (black hole), 104, 148, 151
Magellan Radar Mapper, 29, *29*, 30
Magellanic Clouds, 86, 87, 148

Magnetic field, *26*, 95, 98, *98*, 99, 102, *102*, *103*
Magnetosphere, 28, 59
Malin, David, 85, *85*
Manhattan Project, 143
Maran, Steve, 165
Marcy, Geoff, 77
Mariner 4 (probe), 37
Mariner 10 (probe), 27
Mars, 15, 21, 22, *22*, 27, 34, 35, *35*, 36, *36*, 37, *37*, 38, *38*, 39, *39*, 64, 65, *65*, 66, 133, 134, *134–135*, 135, 136, 138
Mars Climate Orbiter, 37
Mars Direct, 136
Mars Exploration Rover Mission, 35, 38–39, *38–39*
Mars Ghoul, 37
Mars Global Surveyor, 34, 36, *36*, 38
Mars Lander, 134–35
Mars Odyssey, 34, *38*
Mars Pathfinder (rover), 37
Mars Reference Mission, 135–36
Mars Society, 134
Maser light, 90
Matter/antimatter annihilation, propulsion, 144, *145*, 146
Max Planck Institute, 152
Maxwell Montes (Venus), 28
Mayor, Michel, 77
Mazur, Pawel O., 107
MCG 63015 (galaxy), *98*
McKay, Chris, 34
McKay, David, 65
Mediciean moons, 45
Mercury, 22, *22*, 27, 28
Meridiani Planum (Mars), 35, *35*, 39, *39*
Merope (star), 90, 91, *91*
Messenger (probe), 27
Meteors, 32, *32*, 156
M4 (globular star cluster), 80–81
Microbes, 47, 61–65
Microfossils, 65, *65*
Microwave background, 16–17, *17*, 18
Milky Way, 81, 86, 87, 92, 96, 104, 105, *105*, 109, 110, *112*, 113, *113*, 116, *117*, 118, *122*, 123, 164
Millisecond pulsar, 80, 81, 94, *94*
Miranda (Uranian moon), 50
Moon, 12, 15, 21, 22, *22*, 31, 32, *33*, 66, *66*, 73, *126*, 127, 138
Moonlets, 31, 32, *33*, 41, 34, 53
Mossbauer Spectrometer, 39
Mottola, Emil, 107

N
NASA, 27, 29, 34, 41–42, 52, 58, 68, 78, 125, 128, 135, 137, 139, 147, 163
National Academy of Sciences (Washington), 118, 164
NEAR (probe), 41
Nebulae, spiral, 19, 118
Nebular theory, 22–24
Neptune, 49, 50
Neutronium, 94
New Horizons (mission), 52, 139, 141
New Millennium Program (NASA), 143
Newton, Sir Isaac, 22, 120, 151
NGC 4261 (black hole), 96–97, *97*, 101
Nuclear fusion, 23, 27, 43, 80, 84
Nulling interferometry, 78–79

O
O'Dell, Robert, 24, 84–85, 91
Oemler, Gus, 153
Oligotrophic organisms, 64
Oort Cloud, 54, *54*, 55, 110, 124, 144
Oort Cloud Object, 53, *53*
Oort, Jan Henrik, 55
Opportunity (rover), 34, 37, 38–39
Optical wavelengths, 78–79
Orbiter (Mars), 35
Orion Nebula, 24, 84–86, *84*, *85*, 87, *111*, 144, *145*, 148
Orion's Belt, 85
Owen, Tobias, 47

P
Parsec, 110
Particle accelerators, 19
Pegasus (constellation), 77
Penzias, Arno, 17, 18, *18*
Phobos (Martian moon), 40, *40*
Piazzi, Giuseppe, 40
Pillars of Creation (Eagle Nebula), 81, 82, *82*, 84
Pioneer 10 (probe), 55
Planetar, 80
Planetary motion, laws of, 78
Planetary Society, 141
Planet(s), 21, 22, *22*, 24, 75, 77–80
Planck mass, 105–6
Plate tectonics, 28–29
Pleiades, 90, 91, *91*, 148
Pluto, 50, *51*, 52, 110
Positrons, 96
Primordial disk, 90, 91, *91*
Primordial Kernel, 159, *159*
Prism, 120, *121*

Project Orion, 143–44
Proplyds, 24, 84–86, *84*, *85*, 90, 91
Propulsion technologies, 139–46
Protons, 79, 100
Protostars, 80
Proxima Centauri, 23, 84
Ptolemy, 21, 22
Pulsar B16, 20–26, 80–81
Pulsars, 95
Punctuated evolution, 31

Q
Quaoar (KBO), 53, 55
Quarks, 100
Quasars, 104, 156
Queloz, Didier, 77
Quintessence, 163

R
Radiation, 26, 67, 81, 84, 95, 101, 143, 122, 156
Radio communication, 18, 52, 70, 71, 74, 104, 147
Reagan, Ronald, 75
Red dwarf star, 80
Redshift, 18, 118–19, 121
Rees, Sir Martin, 104
Reflection nebulae, 90, 91, *91*
Refregier, Alexandre, 155
Relativity, theory of, 19, 100, 101, 119
Retrograde rotation, 28
Reynolds, Chris, 98
Rock ALH 84001, 65–66, *65*
R–136 (stars), 86, 87
Rubin, Vera, 151
Rutan, Burt, 137

S
Sagan, Carl, 60, 70, 137
Sagittarius A West (black hole), 96, 104, 105, *105*, 113, *113*
Sagittarius A East, 96
Sagittarius Dwarf Spheroidal (galaxy), 116
Sailing technology, 138–141, *140*
Saturn, 22, *22*, 48, *48*, 49, *49*
Scale of the Universe (debate), 118
Schechter, Paul, 153
Schiaparelli, Giovanni, 34
Schmidt, Harrison, 127
Schwarzschild, Karl, 101
Scorpius (constellation), 95
Scowen, Paul, 81, 86

Sedna (KBO), 53, *53*
Semiotics, 71
Serpens (constellation), 81
SETI (Search for Extra Terrestrial Intelligence), 58, 73
Seyfert galaxies, 104
Shapley, Harlow, 86, 118, 162
Shectman, Steve, 153
Shenzhou-5 (spacecraft), 132–33, *132*
Short-period comets, 52
Shuttleworth, Mark, 137
Singularity (black holes), 97, 99, 100, 102, *102, 103*
SIRTF (*Spitzer Space Telescope*), 119, 147
Smith, Gerard, 144, 148
SOHO (*Solar and Heliospheric Observatory*), 25, *26*
Solar flares, 26
Solar sailing, 139–40, *140*, 141, 143
Solar storms, 26
Solar system, 21, 22, *22*, 23, 24, 31, 32, *32–33, 76*, 77, 110
Solar winds, 23, 29, 41, 84, 86, 90, 114, 139
Soviet space program, 28, 29, 37, 132
Soyuz (space capsule), 132, 137
Space Interferometry Mission (*SIM*), 147
Space Ship One, 137, *137*
Space Telescope Science Institute, 123, 160
Spectroscopy, 120–121, *120–21*
Spielberg, Steven, 70
Spiral arms, 110, 111, *111*
Spiral galaxies, 110, 125
Spiral nebulae, 19, 118
Spirit (rover), 37, 38–39, *38–39*, 66
Split-mission approach, 136
Squyres, Steven, 35
Starburst, 87, 124, *124*
Stars, 16, 22–23, 24, 75, 77, 78, 79, 80, 81, 84–87, *84, 88*, 89–90, 91, *91*, 92, *93*, 94–95, *94*, 99, *108*, 109–10, 111, *111*, 113, 114, *115*, 116, 123, 124, 156, 158
Stardust Sample Return (mission), 42
Star Trek, 60, 141
Steinmetz, Matthias, 123
Stephan's Quintet, 124–25, *124*
Stellar winds, 95
Sub-brown dwarf star, 80
Summers, Frank, 123
Sun, 12, 21, 22, *22*, 24, 26, *26*, 27, 101, *122*, 123

Sunspots, 26
Supercluster (Virgo), 149, *149*
Supergalactic plane, 151
Supermassive black holes (SMBHs), 104, 106, 113, *113*, 114, 123, 158
Supernova, 22–23, 24, 89, 90, 92, 94, *94*, 96, 99, 123
Surveyor II (probe), 66
Sword of Orion, 85, *85*

T

Taikonaut, 132, *132*
Tau Ceti (star), 74
Taurus-Littrow region (Moon), 127
Telescopes, new technology, 146–47
Terraforming, 34, 136
Terrestrial Planet Finder (*TPF*), 78–79, 147
Tharsis region (Mars), 36, *36*
30 Doradus, 86–87
Tidal tail, 116, 123, 125
Time, Arrow of, 157, *157*
Titan (Saturnian moon), 48
Tito, Dennis, 137
Trapezium stars (Orion Nebula), 86
Triangulum, 117, *117*
Tunkguska River (Siberia), 42–43
2001: A Space Odyssey, 131
Tyson, Neil, 52

U

UFOs, 56, 69, 72, 73
Uhuru (satellite), 101
Ulam, Stanislaw, 143
Ultraviolet astronomy, 120, *120*, 147
United States, space program, 37, 128, 131–32, 137
Universe, 16, 19, 21, 22, *22*, 107, 118, 152, *152*, 153, *156*, 157, *157*, 158, 159–60, *161*, 162, 163
Uranus, 49, 50
Ursa Major (constellation), 119

V

Valles Marineris (Mars), 135
van Leeuwenhoek, Antonius, 62
Varuna (KBO), 53, 55
Venera (spacecraft), 28, 29
Venus, 15, 22, *22*, 27, 28–30, *29*, 59–60, 64
Very Rapid Burster (star system), 95
Viking landers, 37
Viking/Orbiter (mission), 40

Virgo Cluster, 148, 149, *149*, 151
Volcanism, 45, 59, 67, 135
von Braun, Wernher, 136
von Däniken, Erich, 73
von Fraunhofer, Joseph, 120
von Soldner, Johann George, 154
Vostok, Lake (Antarctica), 64, *64*
Voyager 1, 45, 55
Voyager 2, 48, 50

W

Webb, James, 147
Wegener, Alfred, 28
Welles, Orson, 60
Whirlpool Galaxy, 111, *111*
Wide Field and Planetary Camera 2, 81, 125
Wilms, Joern, 98
Wilson, Robert, 17, 18, *18*
WIMPs (Weakly Interactive Massive Particles), 151, 153
WMAP (*Wilkinson Microwave Anisotropy Probe*), 16–17, *17*, 19, 152, 158, 160, 163

X

XMM X-ray satellite, 98, *98*, 104
X Prize, 137

Y

Yang Liwei, 132, 133
Yellow dwarf star, 24, 92
Yucatan Peninsula, 30

Z

Zheng, Admiral He, 131–32
Zhu Di, emperor of China, 131, 133
Zubrin, Robert, 134–36
Zwicky, Fritz, 155

Image Credits |

Every effort has been made to correctly attribute all material reproduced in this book. If any errors have unwittingly occurred, we will be happy to correct them in future editions.

All art and diagrams, unless otherwise designated, are by Dana Berry © 2004.

ACS	Advance Camera for Surveys
AURA	Association of Universities for Research in Astronomy
AUI	Associated Universities, Inc.
ESA	European Space Agency
CXC	Chandra X-ray Observatory Center
FOC	Faint Object Camera
GSFC	Goddard Space Flight Center
ITSS	Information Technology and Scientific Services
JHU	Johns Hopkins University
JPL	Jet Propulsion Laboratory
MGS	Mars Global Surveyor
MOLA	Mars Orbiter Laser Altimeter Science Investigation
NAOJ	National Astronomical Observatory of Japan
NASA	National Aeronautics and Space Administration
NOAO	National Optical Astronomy Observatory
NRAO	National Radio Astronomy Observatory
NSF	National Science Foundation
NSSDC	National Space Science Data Center
SOHO	Solar and Heliospheric Observatory
SSI	Space Science Institute
STScI	Space Telescope Science Institute
WMAP	Wilkinson Microwave Anisotropy Probe

Front cover: T.A. Rector (NRAO/AUI/NSF and NOAO/AURA/NSF) and B.A. Wolpa (NOAO/AURA/NSF)

Back cover: ESA/NASA and Albert Zijlstra

(Inset, top): NASA, Holland Ford (JHU), the ACS Science Team and ESA

(Inset, middle): T.A. Rector (NOAO/AURA/NSF) and Hubble Heritage Team (STScI/AURA/NASA)

(Inset, bottom): ESA and Garrelt Mellema (Leiden University, the Netherlands)

1: NASA/Digital Stock

2: Subaru Telescope, NAOJ

4: T.A. Rector (NOAO/AURA/NSF) and Hubble Heritage Team (STScI/AURA/NASA)

6: NASA/Digital Stock

10–11: Getty Images

13: Getty Images

CHAPTER ONE

14: NASA/Digital Stock

17: Courtesy of Chuck Bennett and the WMAP Team, NASA/GSFC

18: Roger Ressmeyer/CORBIS/ Magma Photo

CHAPTER TWO

20: Earth Imaging/Getty Images

23: Nolan Walborn, R. Barba, NASA/AURA/STScI Hubble Heritage Team

25: SOHO/GSFC/NASA/ESA

26: SOHO

29: NASA/JPL/Magellan

35: Created using data from NASA/JPL/MGS/MOLA Science Team

36: NASA/JPL/Malin Space Systems (colorization of top image by SkyWorks Digital)

37: NASA/MGS

38: NASA/JPL

39: (top) NASA/ASU/Cornell; (inset) NASA/JPL; (bottom) NASA/JPL/Cornell/USGS

40: (top) Image mosaic, Edwin V. Bell, NSSDC/Raytheon ITSS, NASA/JPL; (bottom) NASA/JPL

42: NASA/JPL

44: John Spencer (Lowell Observatory), NASA/STScI

46: NASA/JPL

47: Based on an original graphic by NASA/JPL

48: NASA/JPL

49: NASA/JPL/SSI

53: (top) NASA/JPL-Caltech; (above) NASA/Caltech

CHAPTER THREE

59: JPL/NASA

62: SkyWorks/Wright Center/Tufts University/E. Chaisson

64: NASA/Marshall Space Flight Center/Russian Academy of Sciences

65: NASA/Johnson Space Center/Stanford University

66: NASA/Johnson Space Center

71: Dale O'Dell/CORBIS/Magma Photo

CHAPTER FOUR

82: Jeff Hester and Paul Scowen, NASA/AURA/STScI

83: T. A. Rector, NOAO/AURA/NSF

84: Robert O'Dell/NASA/AURA/STScI

85: Anglo-Australian Observatory, David

Malin Images

89: NASA/STScI (colorization by SkyWorks Digital)

91: (inset) NASA Hubble HeritageTeam, George Herbig and Theodore Simon (University of Hawaii)

92: Andrew Fruchter, NASA/AURA/STScI

93: NASA/Harvard-Smithsonian CX

97: Walter Jaffe, Holland Ford, NASA/AURA/STScI

98: (inset) John Biretta, Mario Livio, NASA/ESA/AURA/STScI

105: NASA/CXC/MIT/F.K. Baganoff

106: Courtesy of CERN

CHAPTER FIVE

111: NASA/AURA/STScI Hubble Heritage Team, N. Scoville (Caltech) and T.A. Rector (NOAO)

113: NASA/CXC/UMass/D. Wang et al.

115: M. Shara, D. Zurek, L. Drissen, F. Ferraro, B. Paltrinieri, R. Rood and B.Dorman/NASA/AURA/STScI Hubble Heritage Team

119: (left) NASA/JPL-Caltech/K.Gordon (University of Arizona) and S. Willner (Harvard-Smithsonian CfA); (right) Jason Ware

120: Courtesy of ThinkQuest

121: Clayton J. Price/CORBIS/ Magma Photo

124: NASA/AURA, Jayanne English (University of Manitoba), Sally Hunsberger, Sarah Gallagher, Jane Charlton (Pennsylvania State University) and Zolt Levay (STScI)

CHAPTER SIX

126: Neil Armstrong/NASA

129: NASA/Digital Stock

130: NASA

132: Courtesy of Xinhua News Agency

134–5: NASA/JPL/Malin Space Systems (colorization by SkyWorks Digital)

137: Courtesy of Scaled Composites

142: NASA/JPL

145: Orion Nebula backdrop courtesy of Robert O'Dell (Rice University), NASA/AURA/STScI

CHAPTER SEVEN

153: NASA/ESA/STScI/FOC

154: NASA, Andrew Fruchter and the ERO Team [Sylvia Baggett (STScI), Richard Hook (ST-ECF), Zoltan Levay (STScI)], (STScI)

158: (for WMAP image) C. Bennett/ WMAP/NASA; (for Hubble Ultra Deep Field image) Steve Beckwith, NASA/ESA/STScI

EPILOGUE

164: NASA/ESA Hubble Space Telescope

166: NASA, Hubble Heritage Team (AURA/STScI) and ESA

168–9: ESA, NASA and Martino Romaniello (European Southern Observatory, Germany)

Web Sites |

Excellent resources exist on the World Wide Web for fun and further study. The reader is encouraged to visit the following:

Astrobiology Magazine
www.astrobio.net

Chandra X-ray Observatory Center
www.chandra.harvard.edu

European Space Agency
www.esa.int

Jet Propulsion Laboratory
www.jpl.nasa.gov

National Aeronautics and Space Administration
www.nasa.gov

National Optical Astronomy Observatory
www.noao.edu

National Radio Astronomy Observatory
www.nrao.edu

SETI Institute
www.seti.org

Solar and Heliospheric Observatory
sohowww.nascom.nasa.gov

Space.com
www.space.com

Space Science Institute
www.spacescience.org

Space Telescope Science Institute
www.stsci.edu

With Thanks |

Writing this book has taught me one thing — not all the stars are in the sky. Some of the brightest can be found right here on Earth. It was by the light of these stars that this book was written.

A particularly bright cluster of these stars can be found inside the good offices of Madison Press Books in Toronto, Canada. They include Hugh Brewster, whose vision and faith initiated this book, and Ian Coutts, who provided critical guidance to me during the outline and first draft. The art star is Gord Sibley, whose spectacular layouts and graphic design grace these pages. And there is Wanda Nowakowska, a star among stars. Her patience, enthusiasm, keen editor's eye and management skills whipped this book into shape — but the thing I thank her for most is keeping this project fun.

Several bright stars to the south were also extremely influential in the formation of this book, and their expertise provided the guidance necessary to keep this book on target. However, the contribution of Ray Villard, of the Space Telescope Science Institute in Baltimore, Maryland, goes far beyond this book — for it was his invitation to work in the public affairs office at the Institute that set the course for my life's journey among the stars. In close proximity is another star, Jeff Hayes at NASA Headquarters in Washington, D.C. Officially, Jeff is to be thanked for correcting the manuscript, but it is our conversations in Westwood, Pasadena and Dupont Circle that provided much of the substance of this book. Toward the east, at the BBC Open University in London, is Carole Haswell. Carole knows when the hyperbole is too thick, as she so kindly pointed out to me.

In Washington, D.C., there is another cluster of bright stars. These include Don Fehr and Scott Mahler at Smithsonian Books, whose belief in this project helped make it a reality. And two expert stars who read the book on the Smithsonian's behalf — Ted Maxwell, who is really more of a planet guy than a star, and John Huchra, who is more akin to large-scale structure than to any single star!

And speaking of large-scale structure, a special thanks goes to the supergiant star, Eric Chaisson, at Tufts University in Boston, Massachusetts. The beauty of the forward he wrote for this book will be self-evident to the reader, but his mentoring role in the progress of my career and my own comprehension of the cosmos cannot be overstated. His big-picture view of the cosmos provides much of the philosophical backdrop for this book — although his kind advice, "Quote only the dead," wasn't always heeded.

Finally, for me the brightest stars in my firmament are those of my family and friends. I'm especially grateful to my wife, Alla Savranskaia, whose faith and encouragement led to the development of this book. I cannot thank her enough for her patience and forbearance during my many late nights at the computer, and for her help in correcting the manuscript.

It is truly my good fortune to have so many fine people believe in me.

— Dana Berry
Los Angeles

Editorial Director
WANDA NOWAKOWSKA

Art Director
GORD SIBLEY

Manuscript Editor
IAN R. COUTTS

Production Manager
SANDRA L. HALL

Editorial Assistance
IMOINDA ROMAIN, DONNA CHONG

SMITHSONIAN
INTIMATE GUIDE TO THE COSMOS
was produced by
MADISON PRESS BOOKS

MADISON

President and CEO
BRIAN SOYE

Chairman
ALBERT E. CUMMINGS

*Vice President, Business Affairs
and Production*
SUSAN BARRABLE

Publisher
OLIVER SALZMANN

*Director, Business Development
Custom and Contract Publishing*
CHRISTOPHER JACKSON

Managing Editor
IMOINDA ROMAIN

Graphic Designer
JENNIFER LUM

Production Manager
DONNA CHONG

Manager of Finance
GREG THOROGOOD

Administrative Coordinator
TARA KLEIN